Hybrid Enhanced Oil Recovery using Smart Waterflooding

Hybrid Enhanced Oil Recovery using Smart Waterflooding

KUN SANG LEE
Professor
Department of Earth Resources and Environmental Engineering
Hanyang University
Seoul, South Korea

JI HO LEE
Post-Doctoral Researcher
Department of Earth Resources and Environmental Engineering
Hanyang University
Seoul, South Korea

Gulf Professional Publishing
An imprint of Elsevier

HYBRID ENHANCED OIL RECOVERY USING SMART WATERFLOODING ISBN: 978-0-12-816776-2

Publisher: Brian Romer
Senior Acquisition Editor: Katie Hammon
Editorial Project Manager: Ali Afzal-Khan
Project Manager: Poulouse Joseph
Cover Designer: Alan Studholme

50 Hampshire Street, 5th Floor, Cambridge, MA 02139, United States

The Boulevard, Langford Lane, Kidlington, Oxford, OX5 1GB, United Kingdom

Preface

During recent years, systematic effort has established the scientific basis for low-salinity and smart waterflood technology. Moreover, various enhanced oil recovery (EOR) techniques are combined with low-salinity and smart waterflood technology. Smart water-based hybrid EOR techniques are still new and developing technologies. Though widespread successful applications remain at the laboratory and field use is generally limited to pilot scale, the advance of technology is fostering the acceptance of hybrid EOR technologies coupled with smart waterflood through the petroleum community. The applicability of smart water-based EOR technologies, its economic feasibility, and indications for future directions have become essential elements of current EOR research.

The authors believe that more comprehensive reference is needed to facilitate the exchange of information and a discussion of ideas for the field application and future research emphasis. In view of this, the authors prepared this book, which reviews and summarizes engineering fundamentals of smart waterflood coupled with EOR and current progress in research and practical applications. The intent of this book is to provide a rather in-depth review and to be a guide to the engineering aspects of smart waterflood-based EOR technologies.

Chapters 1–3 focus on various aspects of low-salinity and smart waterflood. Chapter 1 serves as an introduction to the topic by tracing the development of low-salinity and smart waterflood technologies in the laboratory and field. Chapter 2 reviews proposed mechanisms in sandstone and carbonate reservoirs. Chapter 3 discusses modeling methods including empirical modeling and mechanistic modeling using geochemistry.

Chapters 4–6 focus on various hybrid EOR technologies combined with low-salinity and smart waterflood. Chapter 4 describes hybrid chemical EOR including polymer flood/gel treatment, surfactant flood, alkaline flood, surfactant-polymer flood, and alkaline-surfactant-polymer flood. Chapter 5 presents hybrid CO_2 EOR including CO_2 WAG and carbonated waterflood. Chapter 6 explains hybrid thermal recovery including hot water injection and steam injection for heavy oil recovery.

This book would never have been published without the able assistance of the Elsevier staff for their patience and excellent editing job. We shall appreciate any comments and suggestions.

Kun Sang Lee
Seoul, Korea

Contents

CHAPTER 1

History of Low-Salinity and Smart Waterflood

ABSTRACT

Historically, the significant investigations regarding the reservoir wettability develop the technology of low-salinity and smart waterflood. Because of the different conditions and the difficult consistency of experiments, many laboratories show a variety of responses of the low-salinity waterflood (LSWF). Especially, the totally different conditions between the sandstone and carbonate rocks hardly draw the universal and consistent results of LSWF in sandstone and carbonate reservoirs. In addition, the increasing oil recovery from LSWF experiments hardly guarantees the successful field deployments of LSWF. Despite the various observations and uncertainty, extensive research studies have clearly observed the enhanced oil production of LSWF in some conditions and agreed the potential of LSWF as enhanced oil recovery technology. Therefore, this chapter reviews the important laboratory and field studies, up to date, to summarize the evidences and experimental conditions of LSWF.

More than 50 years ago, the observations of the salinity effect on waterflood recovery initiate to investigate the potential of low-salinity waterflood (LSWF) in sandstone. In addition, the unexpected higher oil recovery of seawater injection in the carbonate field leads to the investigations of ionic composition on the wettability of carbonates and triggers the research studies of LSWF or smart waterflood in carbonates. Up to date, the many researchers and industries have explored and developed the novel technologies of LSWF or smart waterflood, which are known as LoSal, SmartWater, Desinger Waterflood, and Advanced Ion Management (AIM). Hereafter, the terminology of LSWF is used as a representative. This chapter illustrates the history of LSWF and describes the important observations of experiments and field applications in sandstone and carbonate rocks, respectively.

LABORATORY EXPERIMENTS

This section focuses on explaining experimental observations in sandstone and carbonate rocks. The description of carbonate rocks follows that of sandstone.

Sandstone

Extensive studies of waterflood have interested in the effects of salinity on oil recovery from sandstones and developed the LSWF to improve oil production. Bernard (1967) flooded freshwater and brines into synthetic and natural water-sensitive cores containing clays and investigated the relative effectiveness of salinity on oil recovery. The study assumed that the fresh brine causes the clay hydration, which contributes to the oil production of freshwater injection. The clay bearing synthetic and natural cores are subjects to the experimental study. The synthetic cores have 2% montmorillonite, which has extremely high surface activity, swelling potential, and exchange capacity. The natural cores are provided from Berea sandstone and outcrop near Wyoming. The Berea sandstone core has relatively less clay concentration of 0.1%, but it exhibits high water sensitivity. Another core from Wyoming has expandable clays of 1.2%. The brines are made by controlling NaCl concentration (0.1%, 0.5%, 10%, and 15%). While the brines and freshwater are flooded into the cores and oil recovery, residual oil saturation and pressure gradient are measured. The experiments observe that the injection of freshwater results in less residual oil saturation as well as higher pressure gradient compared with the injection of brines (Fig. 1.1). The study proposed the two mechanisms to explain the observations. Further experiments of constant injection rate or constant differential pressure are carried out to demonstrate the suggested mechanisms and validated the previous observations of increasing oil recovery and pressure drop. In the experiments, the freshwater injection at constant rate

Hybrid Enhanced Oil Recovery using Smart Waterflooding. https://doi.org/10.1016/B978-0-12-816776-2.00001-5

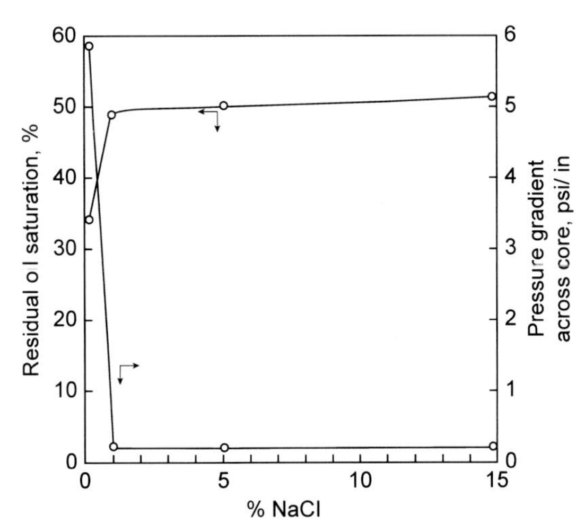

FIG. 1.1 Effects of salinity on the residual oil saturation and pressure gradient. (Credit: From Bernard, G. G. (1967). Effect of floodwater salinity on recovery of oil from cores containing clays. *Paper presented at the SPE California Regional Meeting, Los Angeles, California, USA, 26–27 October*. https://doi.org/10.2118/1725-MS.)

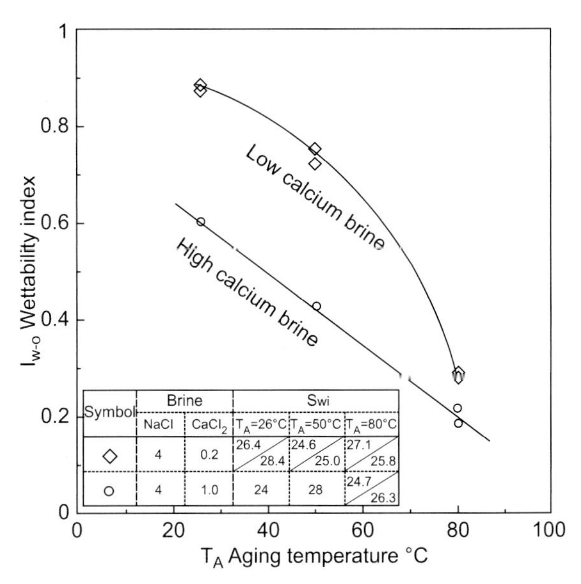

FIG. 1.2 Effects of salinity and aging temperature on wettability index. (Credit: From Jadhunandan, P. P., & Morrow, N. R. (1995). Effect of wettability on waterflood recovery for crude-oil/brine/rock systems. *SPE Reservoir Engineering*, *10*(1), 40–46. https://doi.org/10.2118/22597-PA.)

increases the oil recovery and pressure drop across core. Another fresh water injection at the same pressure differential produces no additional oil. The study concluded that oil recovery increase should be accompanied with the additional pressure drop, and the proposed mechanisms could explain these observations.

Extensive research studies have investigated the relationship between oil recovery and low salinity in terms of wettability. The reservoir wettability is a complex property to determine multiphase flow in porous media and oil recovery of waterflood. Morrow (1990) investigated the wettability of crude oil/brine/rock (COBR) system and its effect on oil recovery of waterflood. Because the accurate understanding and duplicating wettability of reservoir rocks are of importance for numerical simulation and experiment of waterflood, a series of research studies by Morrow and coworkers have tried to quantify the parameters to control the wettability and oil production of waterflood. Jadhunandan and Morrow (1995) reported dominant parameters relating to the wettability of COBR system through experiments. The experiments include wettability index measurements and coreflooding tests of waterflood. Various brines are formulated with NaCl and $CaCl_2$ and tested in the experiments. The brines have NaCl from 4% to 6% and $CaCl_2$ from 0.2% to 2%. Crude oils from West Texas (Moutray) and ST-86, and Berea sandstones are used in the experiments. The wettability index of the Berea core samples is determined by a modified Amott method and spontaneous imbibition experiment. The spontaneous imbibition experiments using the crude oils are carried out by controlling initial water saturation, aging temperature, and salinity. Results of these experiments indicate that wettability index increases when initial water saturation increases and aging temperature decreases. Another experimental results using the Moutray crude oil, not ST-86, report that brine composition and aging temperature change the wettability index (Fig. 1.2). In addition, more than 50 coreflooding tests describe that the close-to-neutral wettability maximizes oil production of waterflood. From these observations, wettability is shown to be sensitive to crude oil type, brine composition, aging temperature, and initial saturation. However, this experimental study hardly reported the effect of brine composition on the wettability and waterflood recovery. Yildiz and Morrow (1996) investigated the potential of brine composition to affect crude oil recovery in coreflooding and spontaneous imbibition tests. It tested the same crude oil from Moutray and two different brines. The Brine 1 is made up of 4% NaCl and 0.5% $CaCl_2$ and Brine 2 had only 2% $CaCl_2$. The experiments investigate various combinations for initial formation and injecting brines using the two types of brine (Brine 1 and Brine 2) and examine the secondary or tertiary recoveries of brine injections. The results of secondary recovery show that injection with Brine 2 into core, which is saturated with Brine 1, increases oil production. However, none of tertiary recovery tests clearly shows the potential of improved oil recovery/enhanced oil recovery (IOR/EOR) by switching brine composition.

Tang and Morrow (1997) proceeded the comprehensive investigations of spontaneous imbibition and waterflooding experiments using Berea sandstone and configured the effects of brine composition, temperature, and crude oil composition on the wettability and oil recovery. Compared with previous studies, this study made a significant effort on the assessments of crude oil and brine compositions in the temperature range from 22°C to 80°C. The synthetic brines and three crude oil samples (Dagang, A-95 of Prudhoe Bay, and CS) are subject to the experiments. In the imbibition and waterflooding tests, the synthetic brines and various diluted versions of the synthetic brines, which have salinities by factors of 0.01, 0.1, and 2, are used. Additional experiments use the modified crude oils in which light ends are removed or alkanes (pentane, hexane, and decane) are added. The summarized results of the tests indicate that both imbibition rate and oil recovery increase with a decreasing salinity. In detail, higher oil recovery is obtained when either connate or invading brines have low salinity (Figs. 1.3 and 1.4). In the assessments of crude oil composition, the existences of light ends and alkanes decrease oil recovery.

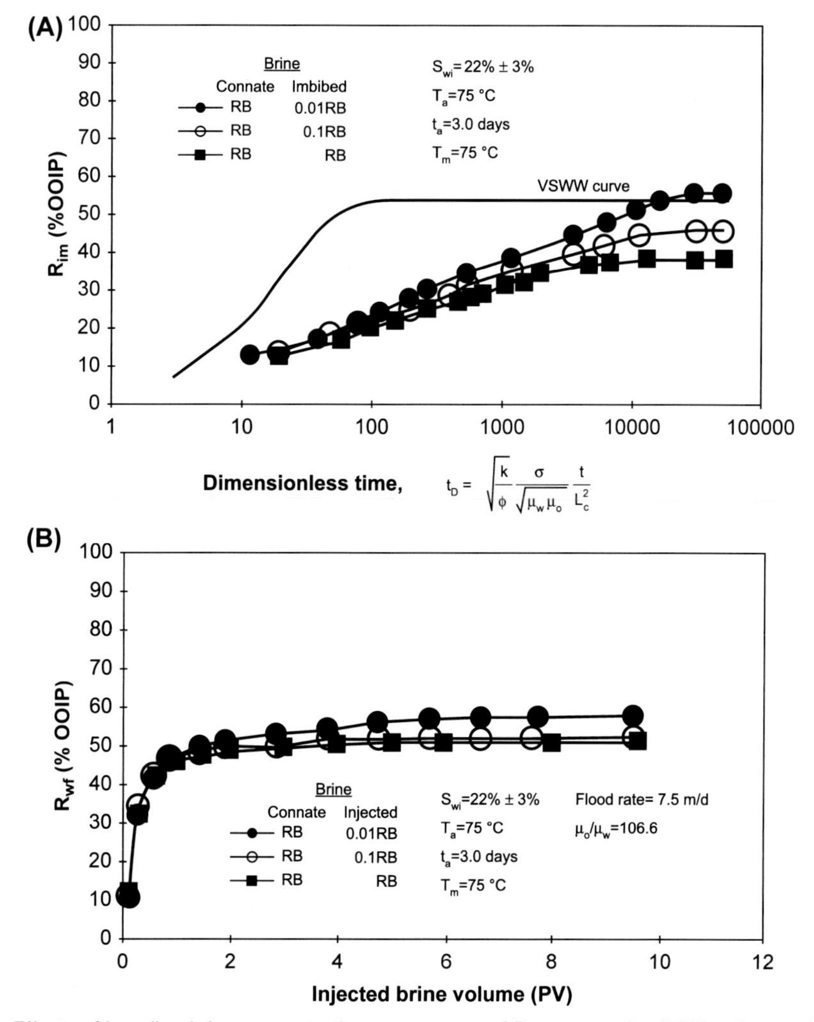

FIG. 1.3 Effects of invading brine concentration on recovery of Dagang crude oil (RB = Dagang brine): **(A)** spontaneous imbibition and **(B)** waterflood. (Credit: From Tang, G. Q., & Morrow, N. R. (1997). Salinity, temperature, oil composition, and oil recovery by waterflooding. *SPE Reservoir Engineering, 12*(04), 269–276. https://doi.org/10.2118/36680-PA.)

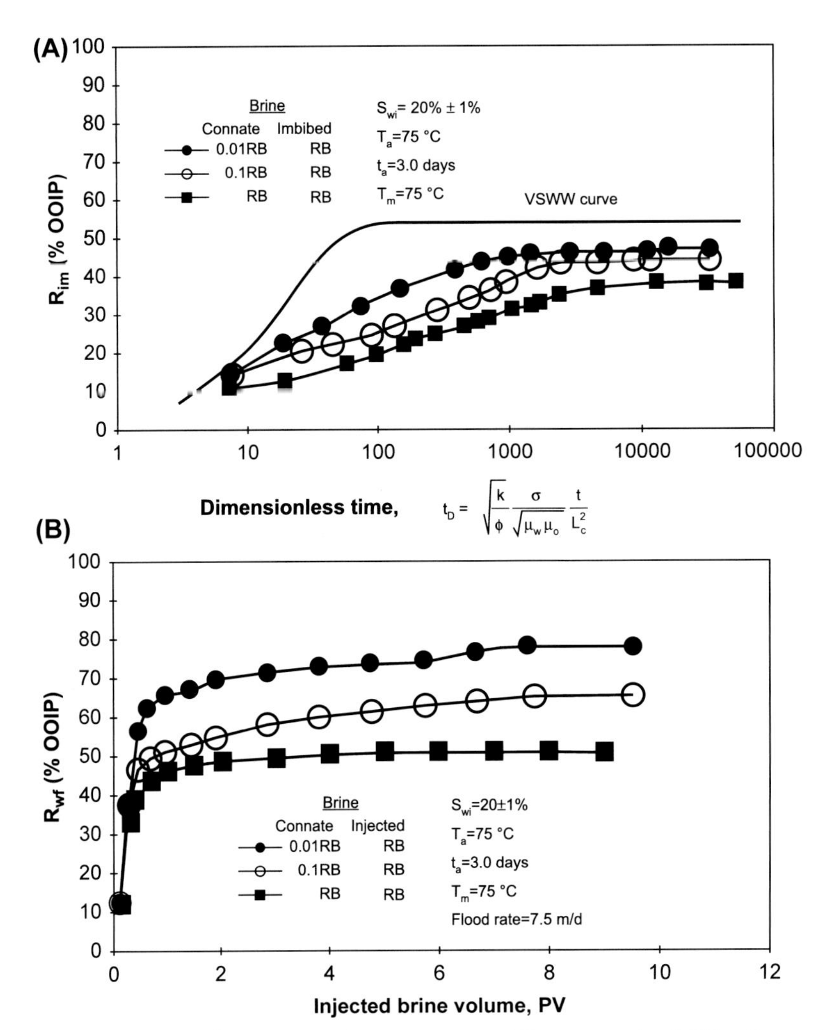

FIG. 1.4 Effects of connate brine concentration on recovery of Dagang crude oil (RB = Dagang brine): **(A)** spontaneous imbibition and **(B)** waterflood. (Credit: From Tang, G. Q., & Morrow, N. R. (1997). Salinity, temperature, oil composition, and oil recovery by waterflooding. *SPE Reservoir Engineering, 12*(04), 269–276. https://doi.org/10.2118/36680-PA.)

This study clearly demonstrated that the low-saline brine has a positive effect on both wettability and waterflood efficiency and oil composition also affects them. Therefore, this study concluded that the wettability is a complex characteristic, which responds to the changes of brine composition, temperature, and oil composition in thermodynamic condition.

Further study (Tang & Morrow, 1999) was continued to reveal how low-saline brine increases the crude oil recovery. In this study, the increasing oil recovery with a decrease in salinity is assumed to be attributed to the fine particle. To verify it, the comprehensive waterflooding and imbibition test are carried out using nontreated and fired/acidized sandstones (Fig. 1.5). Firing and acidizing treatments stabilize the fine particles in cores. In addition, the experiments also test crude and refined oils to validate the effects of crude oil composition, which is observed in the previous study of Tang and Morrow (1997). The sandstone cores of Berea, Bentheim, CS, and Clashach are saturated with synthetic reservoir brine and flooded with various diluted versions of synthetic brine and seawater. The results of waterflooding and imbibition test, using nontreated cores and crude oils, report the oil recovery

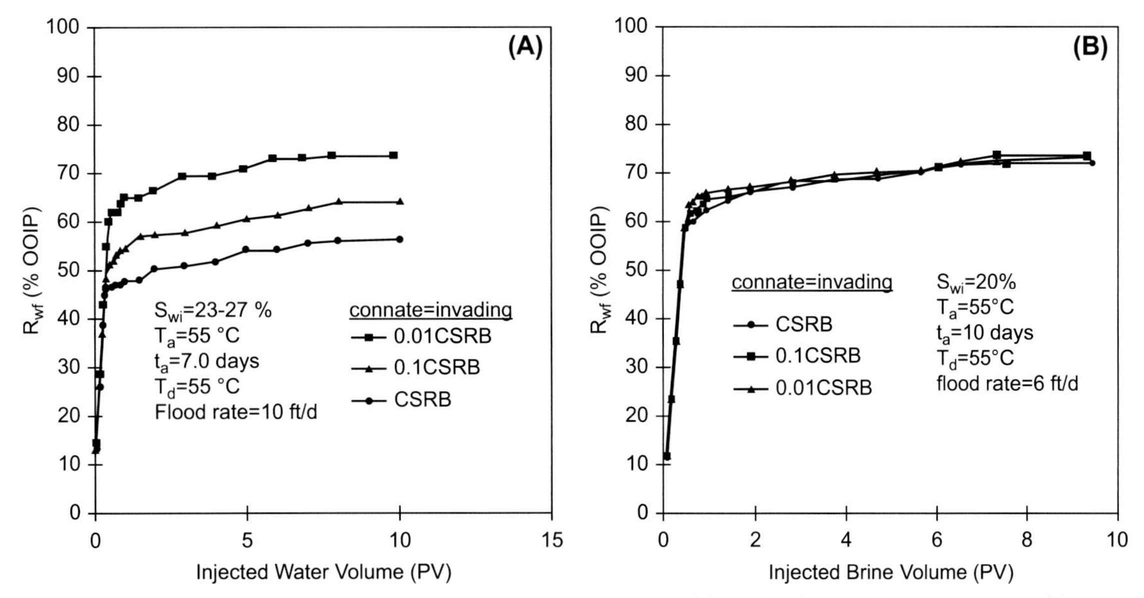

FIG. 1.5 Effects of fine particles on the oil recovery of waterflood: **(A)** nonfired/acidized Berea sandstone, **(B)** fired/acidized Berea sandstone. (Credit: From Tang, G.-Q., & Morrow, N. R. (1999). Influence of brine composition and fines migration on crude oil/brine/rock interactions and oil recovery. *Journal of Petroleum Science and Engineering, 24*(2), 99–111. https://doi.org/10.1016/S0920-4105(99)00034-0.)

increase, when the invading brine has low salinity. However, additional experiments using fired/acidized sandstones or refine oils produce no change in oil recovery. These results indicate that all factors of connate and injection brines, crude oil, and the rock affect the sensitivity of oil recovery to brine composition. Based on these observations, Tang and Morrow (1999) proposed the mechanism of fine migration behind the LSWF.

Agbalaka, Dandekar, Patil, Khataniar, and Hemsath (2008) conducted the coreflooding of LSWF as secondary and tertiary recoveries. They monitored the change of residual oil saturation with variation in wettability, salinity, and temperature. The brines to be tested have salinities of 4%, 2%, and 1%. In the EOR potential test, the experiments switch the injecting brine from high-saline brine to low-saline brine and elevate temperature of injecting brine. They observe that residual oil saturation is reduced from 39% to 15% for decreasing salinity and increasing temperature (Fig. 1.6). Another study by Lager, Webb, Black, Singleton, and Sorbie (2008) also evaluated the potential of LSWF as secondary and tertiary recoveries. The study recorded pH of effluent fluid as well as oil recovery. In addition, it carried out the ion analyses to explain the LSWF in terms of geochemistry. In the ion analyses, the concentrations of divalent cations (Ca^{2+} and Mg^{2+}) between injecting and effluent brines are measured and compared (Fig. 1.7). The concentrations of the effluent brine drop lower than the concentrations in the injecting brine. The observations are explained with adhering Ca^{2+} and Mg^{2+} onto rock matrix. Based on the observations of retardations of Ca^{2+} and Mg^{2+}, a hypothetical mechanism of multicomponent ionic exchange (MIE) is formulated for LSWF.

Ligthelm et al. (2009) conducted the spontaneous imbibition test and coreflooding using Berea and Middle Eastern sandstone cores. They tested various brines including pure NaCl brine, $CaCl_2$ brine, $MgCl_2$, brine, and synthetic brine from Dagang to investigate the role of divalent cations. In the spontaneous imbibition tests, it is found that both pure $CaCl_2$ and $MgCl_2$ generally reduce residual oil saturation less than NaCl brine and the synthetic brine. These findings indicate that the multivalent cations of the brine make the reservoir rock less water-wet. This interpretation is also inferred from the coreflooding. In the coreflooding experiment, the Berea sandstone core to be tested is saturated with 2400 mg/L NaCl brine and Brent Bravo crude oil. This core is flooded by 2400 mg/L NaCl brine following 24,000 mg/L $CaCl_2$ brine. Although there is negligible possibility of formation damage, increasing differential pressure is observed during $CaCl_2$ brine injection. In addition, when the brine injection is changed from $CaCl_2$ brine to NaCl brine, the oil production is resumed despite the differential pressure

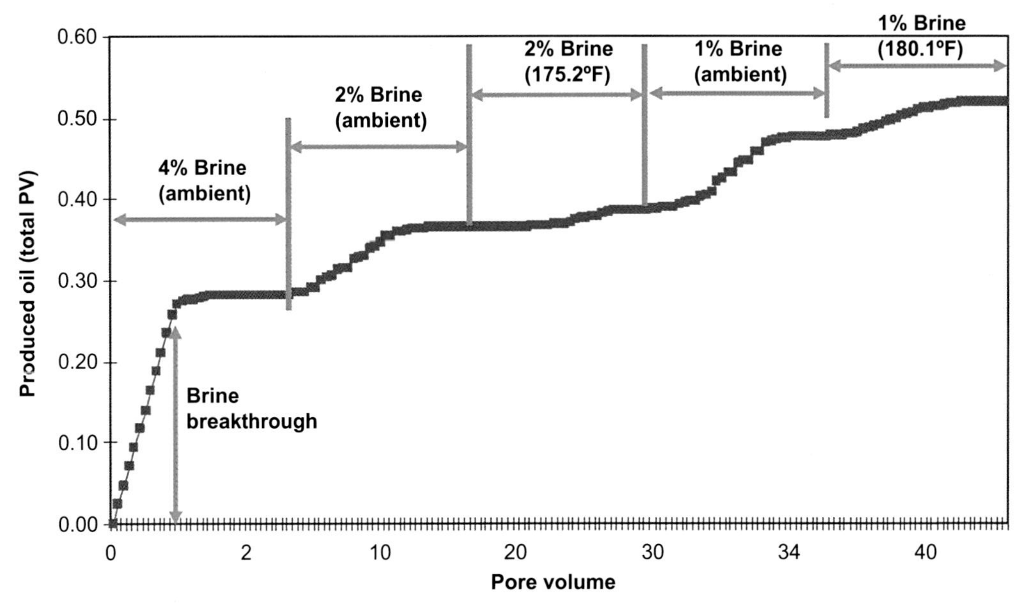

FIG. 1.6 Effects of temperature and salinity on oil production during low-salinity waterflood. (Credit: From Agbalaka, C. C., Dandekar, A. Y., Patil, S. L., Khataniar, S., & Hemsath, J. R. (2008). Coreflooding studies to evaluate the impact of salinity and wettability on oil recovery efficiency. *Transport in Porous Media, 76*, 77–94. https://doi.org/10.1007/s11242-008-9235-7.)

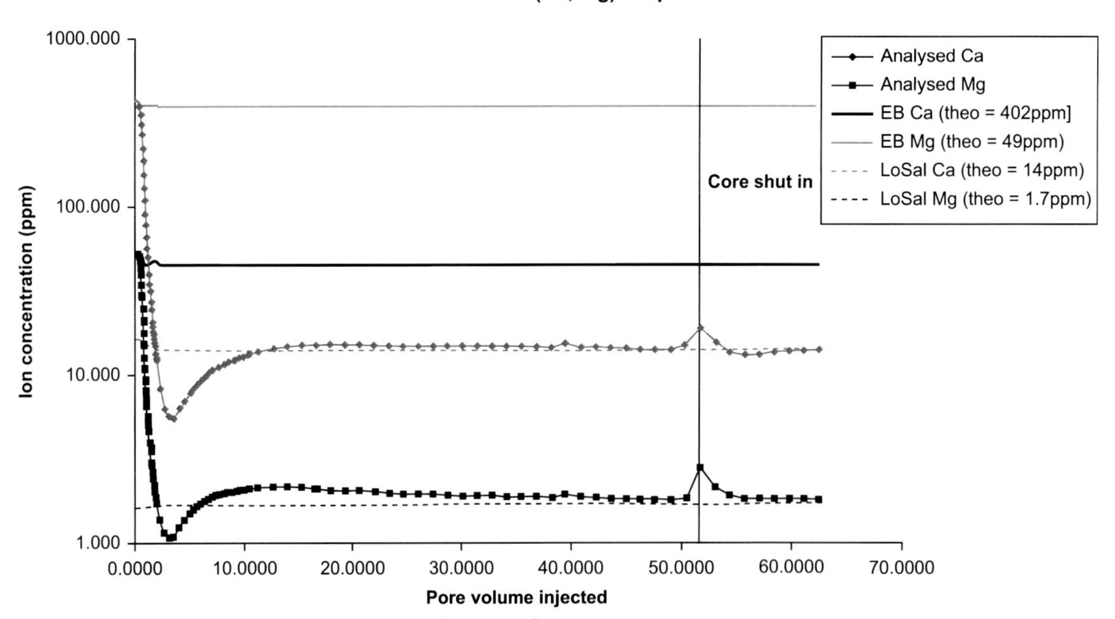

FIG. 1.7 History of concentrations of Ca^{2+} and Mg^{2+} in invading and effluent brines. (Credit: From Lager, A., Webb, K. J., Black, C. J. J., Singleton, M., & Sorbie, K. S. 2008a. Low salinity oil recovery – an experimental investigation1. *Petrophysics, 49*(1), 28–38. https://doi.org/10.2118/93903-MS.)

drop. These observations imply the ability of $CaCl_2$ brine to change reservoir wettability toward more oil-wet. Experiments using Middle Eastern sandstone cores are carried out to review the observations by Lager et al. (2008) and reveal the main factor for LSWF. The core saturated with oil is flushed by formation brine. The formation brine having TDS with 238,000 mg/L is composed of Na^+ with 84,300 mg/L, Ca^{2+} with 6800 mg/L, and Mg^{2+} with 1215 mg/L. The pure NaCl without $CaCl_2$ brines and NaCl with $CaCl_2$ brines are injected in to the core. The two pure NaCl brines are designed with 2000 and 240,000 mg/L and other brines have NaCl with 2000 mg/L and $CaCl_2$ with 10 or 100 mg/L. The injection of pure NaCl with 2000 mg/L produces higher oil production rate as well as the lower level of differential pressure compared with the injections of other brines. Based on these observations of sandstone cores, this study concluded that a major contribution on the increasing oil recovery is the ionic concentration of brine, i.e., ionic strength, rather than the Ca^{2+} and Mg^{2+} and proposed the electrical double layer (EDL) expansion theory as a mechanism of LSWF.

Berg, Cense, Jansen, and Bakker (2010) carried out the experiments to find the direct evidence indicating the exact mechanism of LSWF. They constructed experimental system to visualize the microscopic detachment of crude oil from clay layer. The experiments monitor the movement of oil droplets attached to montmorillonite clay layer as well as thickness of the layer, when salinity of injecting brine is changed from high salinity to low salinity. It is observed that approximately up to 80% of oil is released from the clay layer with the minor swelling of the clay layer. It also reports no deflocculation or release of clay particles. The study concluded that the LSWF increases oil recovery because of wettability modification rather than fine migration and selective plugging via clay swelling.

Austad, Rezaeidoust, and Puntervold (2010) published the adsorption and coreflooding experiments of LSWF to illustrate the effects of pH and salinity. The adsorption measurement uses the kaolinite clay powder, basic organic materials of quinoline, and acidic organic material of 4-tert-butyl benzoic acid. The adsorption of the organic materials on the clay power is measured in the various ranges of salinity and pH conditions. In the low pH condition of 5, the increasing adsorption is observed as salinity decreases. In the high pH condition of 8, the sensitivity of adsorption to the salinity depends on the salinity and the degree of adsorption is relatively low (Fig. 1.8). This observation implies that an increase in pH has much higher impact on the adsorption of organic material than a gradient in the salinity. In coreflooding tests, production, salinity, and effluent pH are monitored. The experimental results indicate that LSWF causes a local increase in pH and the pH increases result in the desorption of organic components of crude oil from the clay. Based on the results, the salting-in mechanism is proposed for LSWF. Another study by RezaeiDoust, Puntervold, and Austad (2011) carried out further experiments to verify the mechanism.

Nasralla and Nasr-El-Din (2014) investigated the LSWF with contact angle measurement, zeta (ζ) potential measurement, and corefloodings of secondary and tertiary oil recoveries. Through the comprehensive experiments, they tried to explain the reason of the improved oil recovery with EDL expansion. Various brines, which have TDS from 109 to 174,156 mg/L and pH from 4 to 7.6, are subject to the experiments. The experiments measure the ζ-potentials of oil/brine and Berea sandstone rock/brine interfaces by changing brine type. The experimental results indicate that ζ-potential of oil/brine is a function of salinity and cation type. In addition, it is found that lower pH produces the less negative and closer to zero ζ-potential. The contact angle is measured to confirm the wettability alteration by changing brine pH and salinity. It is also tried to quantify the relation between the wettability modification and ζ-potential change and develop the relation with the results of LSWF coreflooding. In the test of the tertiary recovery of LSWF, no additional oil recovery and increasing pressure drop due to fine migration are observed (Fig. 1.9A). However, the secondary recovery of LSWF increases oil recovery up to 12% than high-salinity brine injection (Fig. 1.9B). The another coreflooding of LSWF investigates the effects of salinity, brine composition, and pH on the secondary oil recovery. From the experiments, the study concluded that the expansion of EDL is controlled by salinity and pH and it increases the secondary oil recovery when injecting brine has lower salinity and higher pH. The additional study of Shehata and Nasr-El-Din (2017) carried out various experiments including spontaneous imbibition, coreflood, computed tomography (CT) scan, X-ray diffraction (XRD), X-ray fluorescence (XRF), scanning electron microscope (SEM), nuclear magnetic resonance (NMR), and mercury injection capillary pressure (MICP) tests. The extensive experiments intensively investigate the effect of connate water composition on the oil recovery of LSWF and observe that the divalent cations in connate water significantly increase the oil recovery.

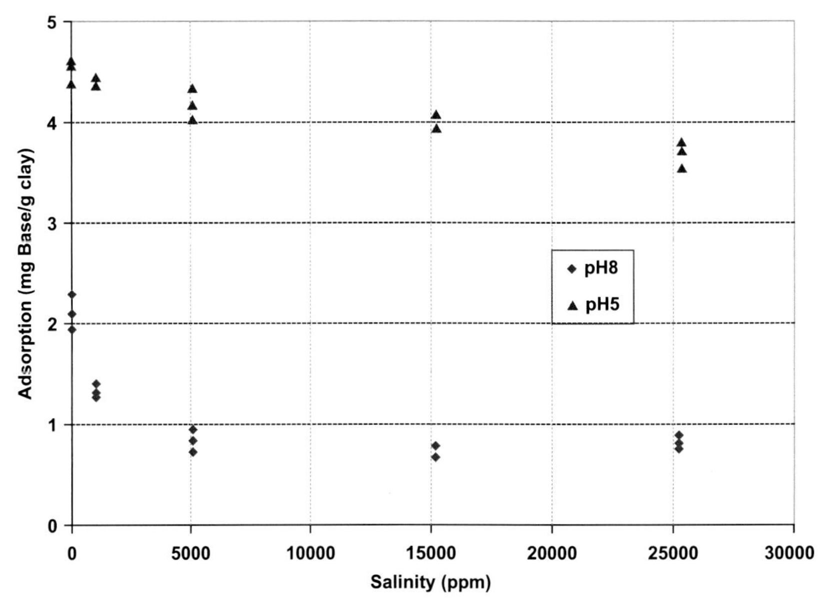

FIG. 1.8 Adsorption of quinolone toward kaolinite clay powder as a function of salinity and pH. (Credit: From Austad, T., Rezaeidoust, A., & Puntervold, T. (2010). Chemical mechanism of low salinity water flooding in sandstone reservoirs. *Paper presented at the SPE improved oil recovery Symposium, Tulsa, Oklahoma, USA, 24–28 April*. https://doi.org/10.2118/129767-MS.)

The previous laboratory-scale experiments have observed the increasing oil recovery of LSWF and analyzed the major factors to affect the oil recovery of LSWF in sandstone reservoirs. The next section thoroughly discusses the experimental observations of LSWF in carbonate rocks.

Carbonate Rocks

Oil production from fractured and low-permeable carbonate is a challenge because the wettability of carbonates is generally neutral to preferentially oil-wet. The unfavorable wettability preventing spontaneous imbibition mitigates water flow from fracture to matrix and oil flow from matrix to fracture and results in the high residual oil saturation. The successful oil production from the carbonates requires wettability modification improving spontaneous imbibition. EOR research group at University of Stavanger has worked with LSWF or smart waterflood in carbonate reservoirs. Austad, Strand, Høgnesen, and Zhang (2005) were interested in the successful injection of seawater into Ekofisk chalk field in the North Sea (Hallenbeck, Sylte, Ebbs, & Thomas, 1991). The Ekofisk is mixed-wet and highly fractured, it has low matrix permeability about 2 mD, and the reservoir temperature is up to 130°C. Hallenbeck et al. (1991) reported a pilot test in Ekofisk chalk field and successful performance of waterflood. The pilot test began in June 1986 and its performance was beyond the expectations of laboratory experiments. Austad et al. (2005) were curious about how seawater injection brought an unexpected success in the Ekofisk chalk. They concluded that the wetting conditions of carbonates are affected by the pH of the brine, temperature of the reservoir, composition of crude oil such as acid/base number (AN/BN), and ionic composition of brine. Therefore, they investigated the effect of temperature, the AN/BN of oil, and synthetic Ekofisk brines on the wettability of carbonates by conducting spontaneous imbibition and chromatographic tests. Especially, the assessments of ionic composition of brine focus on the concentrations of Ca^{2+} and SO_4^{2-}. Generally, the ions of Ca^{2+} and SO_4^{2-} are potential-determining ions to control charge type and density on the chalk surface. Because they influence an adsorption of negatively charged carboxylic components from the crude oil, they have a potential to influence the wetness of carbonates. The results of spontaneous imbibition test illustrate that brines with SO_4^{2-} produce more oil than brines without SO_4^{2-} and more additional oil is recovered for higher concentration of the SO_4^{2-}. The chromatographic results show that the adsorption of SO_4^{2-} on the carbonate surface increases in high temperature condition. These observations predict that brine with high SO_4^{2-} concentration

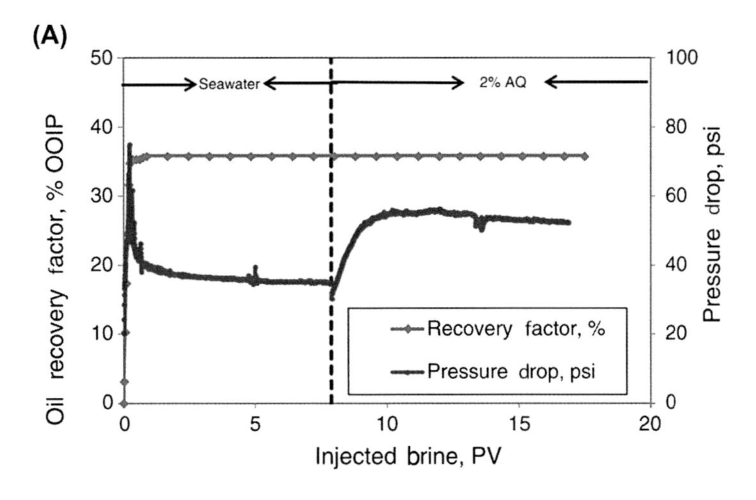

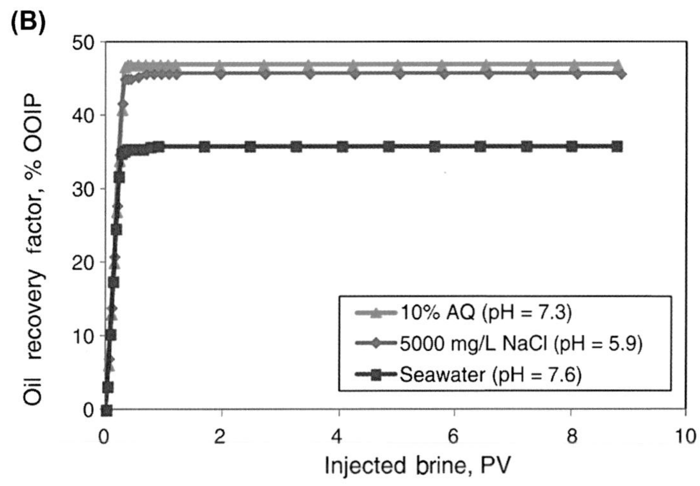

FIG. 1.9 Productions of LSWF as **(A)** tertiary recovery and **(B)** secondary recovery. (Credit: From Nasralla, R. A., & Nasr-El-Din, H. A. (2014). Double-layer expansion: Is it a primary mechanism of improved oil recovery by low-salinity waterflooding? *SPE Reservoir Evaluation & Engineering*, 17(01), 49–59. https://doi.org/10.2118/154334-PA.)

would produce higher oil recovery in high temperature condition. To confirm the prediction, the additional imbibition tests are performed. The oil recovery of the imbibition test is only 15% of original oil in place (OOIP) when imbibing fluid has zero concentration of SO_4^{2-}. The addition of SO_4^{2-} increases the oil production up to 65% of OOIP. These observations explain that the adsorption of SO_4^{2-} onto the chalk surface lowers the positive charge density of chalk and encourages the desorption of negatively charged polar components, i.e., carboxylic group component, of crude oil. In addition, the experiments also analyze the SO_4^{2-} and Ca^{2+} of produced water from Ekofisk chalk. The produced water shows a strong depletion in the concentration of SO_4^{2-} and the relatively higher concentration of Ca^{2+} than the concentration of SO_4^{2-}. These results also indicate that SO_4^{2-} has a strong affinity on the chalk surface remaining strongly positively charged surface of Ekofisk chalk.

Strand, Høgnesen, and Austad (2006) explored the effects of the potential-determining ions (Ca^{2+} and SO_4^{2-}) and temperature on wettability alteration of carbonates in detail. They studied the impact of Ca^{2+} on the adsorption of SO_4^{2-} at various temperature conditions. They carried out spontaneous imbibition, chromatographic, and ζ-potential tests. The spontaneous imbibition test using oil-wet chalk confirms the positive effect of SO_4^{2-} on oil recovery at 110°C. As the brine has the higher concentration of SO_4^{2-}, the oil recovery of the imbibition test also increases.

In the chromatographic studies, the effluent concentrations of Ca^{2+}, SO_4^{2-}, and nonadsorbing tracer thiocyanate (SCN^-) are measured at 23, 70, 100, and 130°C. It is observed that the productions of SO_4^{2-} and Ca^{2+} become more retarded with an increase in temperature (Figs. 1.10 and 1.11). The study concluded that the retardation of SO_4^{2-} is attributed to the adsorption of SO_4^{2-} on the more positively charged chalk surface in a high temperature. In addition, it is expected that the coadsorption of Ca^{2+} by more injection of Ca^{2+} enhances the adsorption of SO_4^{2-}. The ζ-potential measurement tests try to confirm the expectation. Because the high amount of chalk particles leads to low competition between the affinities of Ca^{2+} and SO_4^{2-} toward chalk surface, the measurements fail to draw meaningful observations. Zhang and Austad (2006)

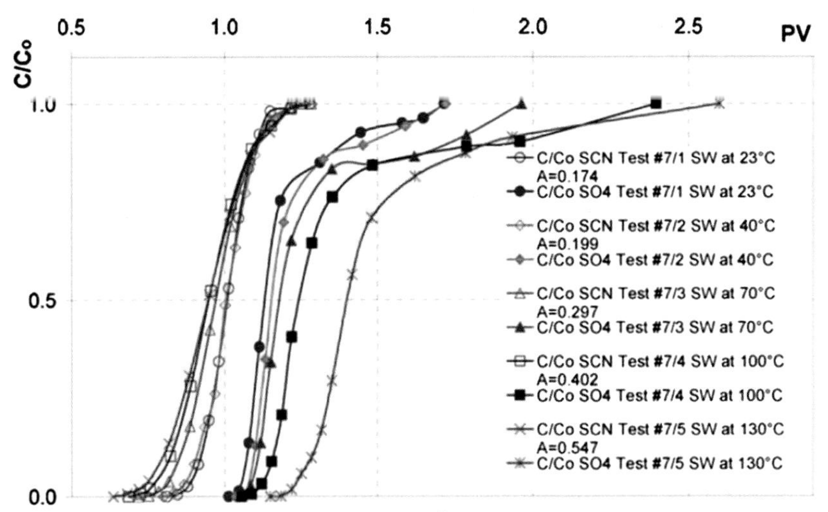

FIG. 1.10 Histories of effluent concentrations of SO_4^{2-} and SCN^{-1} at various temperature conditions. (Credit: From Strand, S., Høgnesen, E. J., & Austad, T. (2006). Wettability alteration of carbonates—effects of potential determining ions (Ca^{2+} and SO42−) and temperature. *Colloids and Surfaces A: Physicochemical and Engineering Aspects*, 275(1), 110. https://doi.org/10.1016/j.colsurfa.2005.10.061.)

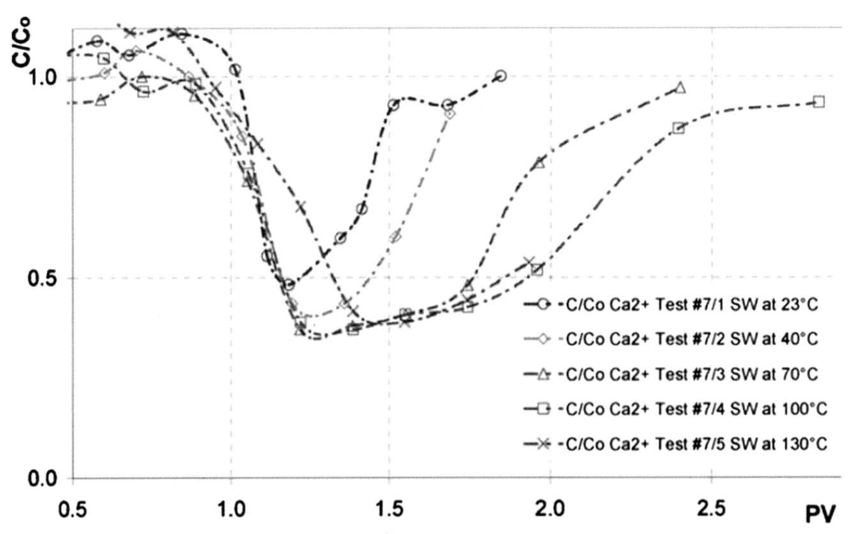

FIG. 1.11 Histories of effluent concentration of Ca^{2+} at various temperature conditions. (Credit: From Strand, S., Høgnesen, E. J., & Austad, T. (2006). Wettability alteration of carbonates—effects of potential determining ions (Ca^{2+} and SO42−) and temperature. *Colloids and Surfaces A: Physicochemical and Engineering Aspects*, 275(1), 110. https://doi.org/10.1016/j.colsurfa.2005.10.061.)

experimentally reported the sensitivity of potential-determining ions, the ratio of SO_4^{2-} to Ca^{2+}, and temperature to the oil recovery of spontaneous imbibition test (Fig. 1.12). They verified the previous observations of Strand et al. (2006) to modify wettability of chalk because of the injection of SO_4^{2-} and high temperature.

Zhang, Tweheyo, and Austad (2007) reported the potential of Mg^{2+} as the potential-determining ion as well as Ca^{2+} and SO_4^{2-}. The study evaluated the impacts of Mg^{2+} present in seawater on spontaneous imbibition, ζ-potential, chromatographic test, and coreflooding. They tested the different brines adjusting concentration of Ca^{2+}, Mg^{2+}, and SO_4^{2-}. The ζ-potential measurements use milled outcrop chalk form Steven Klint near Copenhagen, Denmark. In the tests, the increasing ζ-potential is observed with an increase in the concentration of Mg^{2+}. This observation implies the potential of Mg^{2+} to adjust the charge density on chalk surface. The coreflooding tests at low and high temperature conditions are carried out to confirm the potential of Mg^{2+} according to temperature. In the test, brine containing Ca^{2+}, Mg^{2+}, and SCN^{-1} is flooded into chalk core. The chromatographic test analyzes the effluent concentrations of Ca^{2+}, Mg^{2+}, and SCN^{-1} obtained from the coreflooding. In the low temperature with 23°C, the productions of both Ca^{2+} and Mg^{2+} are delayed compared with the production of SCN^{-} and, especially, there is more delay in the production of Ca^{2+} compared with the production of Mg^{2+}. The result approximately estimates that the Ca^{2+} has 3.4 times higher affinity toward chalk surface over Mg^{2+} in the low temperature. In high temperature with 130°C, the production of Mg^{2+} is seemed to be more retarded than that of Ca^{2+} (Fig. 1.13). In addition, the effluent concentration of Ca^{2+} is higher than the initial concentration of Ca^{2+}. After two pore volume (PV) injection, the effluent concentration of Ca^{2+} decreases to the initial concentration of Ca^{2+}. These observations explained a substitution reaction between Mg^{2+} and Ca^{2+} on the chalk surface. The brine to be used in chromatographic test has the equivalent concentrations of Ca^{2+} and $Mg^{2|+}$. Because the seawater has about four times higher concentration of Mg^{2+} than the concentration of Ca^{2+}, the seawater injection introduces a pronouncing substitution reaction between Mg^{2+} and Ca^{2+}. The another series of spontaneous imbibition tests are designed to configure the interaction between different potential-determining ions of Ca^{2+}, $Mg^{2|+}$, and SO_4^{2-}. The spontaneous imbibition tests vary temperature from 70 to 130°C and concentrations of Ca^{2+}, Mg^{2+}, and SO_4^{2-} (Fig. 1.14). In the tests at 70°C, brines have the zero concentrations of Ca^{2+} and Mg^{2+}, but the concentration of SO_4^2 from zero to four times of seawater. In all brine cases, the only 10% of oil recovery is observed. In the tests at 100°C, no additional oil is observed until the either of Ca^{2+} or Mg^{2+} is added into the brines. These results indicate that SO_4^{2-} without Ca^{2+} or Mg^{2+} has no potential to modify wettability increasing imbibition. When the Ca^{2+} or Mg^{2+} is added in the brines, all cases show the increasing oil recoveries from 20% to 42%.

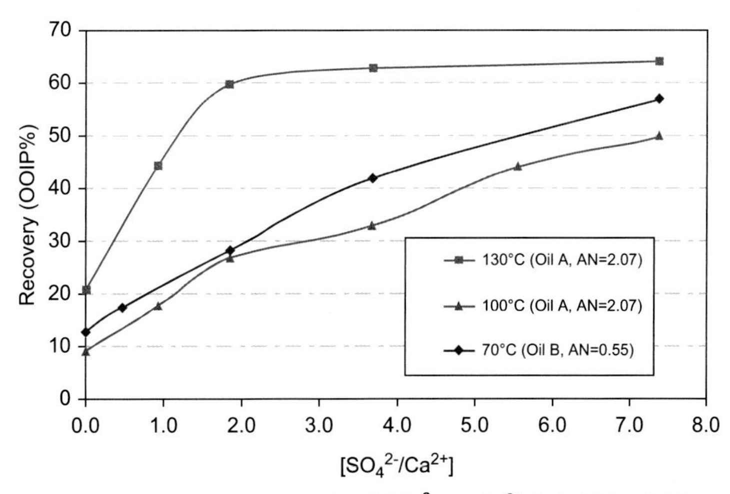

FIG. 1.12 Correlation between oil recovery, ratio of SO_4^{2-} to Ca^{2+} in imbibing fluids, and temperature effects. (Credit: From Zhang, P., & Austad, T. (2006). Wettability and oil recovery from carbonates: Effects of temperature and potential determining ions. *Colloids and Surfaces A: Physicochemical and Engineering Aspects, 279*(1), 179–187. https://doi.org/10.1016/j.colsurfa.2006.01.009.)

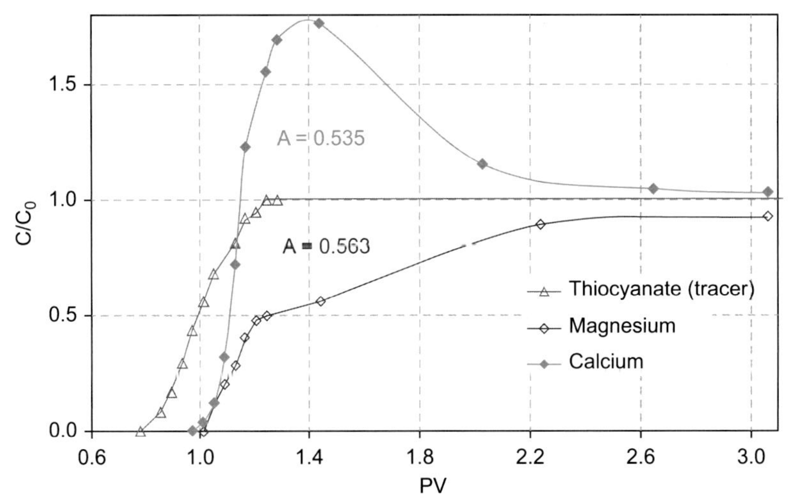

FIG. 1.13 History of effluent concentrations of Ca^{2+}, Mg^{2+}, and SCN^{-1} of coreflooding at 130°C. (Credit: From Zhang, P., Tweheyo, M. T., & Austad, T. (2007). Wettability alteration and improved oil recovery by spontaneous imbibition of seawater into chalk: Impact of the potential determining ions Ca^{2+}, Mg^{2+}, and SO_4^{2-}. *Colloids and Surfaces A: Physicochemical and Engineering Aspects, 301*(1):199–208. https://doi.org/10.1016/j.colsurfa.2006.12.058.)

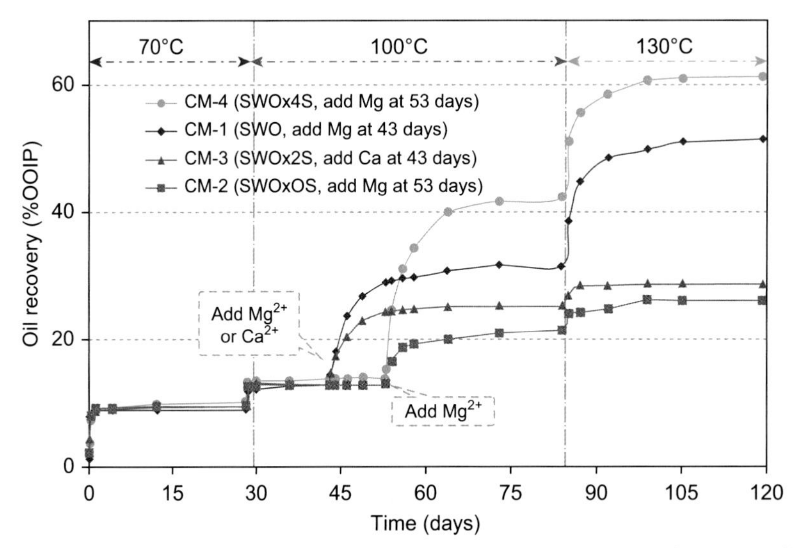

FIG. 1.14 Oil recovery of the spontaneous imbibition tests at various temperatures (70, 100, and 130°C) and various concentrations of Ca^{2+}, $Mg^{2|+}$, and SO_4^{2-}. (Credit: From Zhang, P., Tweheyo, M. T., & Austad, T. (2007). Wettability alteration and improved oil recovery by spontaneous imbibition of seawater into chalk: Impact of the potential determining ions Ca^{2+}, Mg^{2+}, and SO42–. *Colloids and Surfaces A: Physicochemical and Engineering Aspects, 301*(1):199–208. https://doi.org/10.1016/j.colsurfa.2006.12.058.)

In addition, brines with addition of Mg^{2+} produce higher oil recovery with an increase in the concentration of SO_4^{2-}. It implies the strong relationship between Mg^{2+} and SO_4^{2-} on the wettability modification. In the tests at 130°C, all cases increase oil recoveries and brines containing Mg^{2+} and SO_4^{2-} result in oil recovery up to 60%. When the brine only has lack of Mg^{2+} or SO_4^{2-}, the oil recovery is less than 30%. These observations indicate that the efficiency of Mg^{2+} as wettability modifier dramatically increases when brine has SO_4^{2-} and high temperature. This study summarized and concluded that potential of Ca^{2+}, $Mg^{2|+}$, and SO_4^{2-} to adjust the charge density of chalk surface by the adherence of potential-determining ions on the surface modifies wettability of chalk. In addition, Mg^{2+} with SO_4^{2-} has a high affinity on the chalk surface, and the affinity increases in high temperature condition. Based on these conclusions, this study proposed the different mechanisms for the wettability alteration induced by seawater in low and high temperatures, respectively.

The upstream research team from Saudi Aramco has initiated a research program called "SmartWater Flood" to explore the IOR/EOR from carbonates by modifying the brine composition. The research team reported the comprehensive results of laboratory studies to investigate the impact of salinity and ionic composition in COBR system. In the study (Yousef, Al-Saleh, Al-Kaabi, & Al-Jawfi, 2011), the carbonate cores, which is composed of 80% calcite, 13% dolomite, 6% anhydrite, and less than 1% quartz, are subject to the experiments. The field connate water with 213,734 ppm TDS and seawater with 57,670 ppm TDS are used in these experiments. To quantify the impact of ionic composition, the various diluted seawaters by factors of 0.5, 0.1, 0.05, and 0.01 are exploited in the comprehensive experiments. The seawater has more than 10 times higher concentration of SO_4^{2-} over the connate water. Crude oil has 30°API and total acid number (TAN) with 0.25 mg KOH/g oil. This study conducted the IFT measurement, contact angle measurement, coreflooding, and NMR tests. The IFT measurements are carried out using live oil and the various brines at the reservoir temperature with 212°F. The dilution of seawater slightly decreases the IFT, but the reduction of IFT is less than 2 dyne/cm. There is a relatively higher reduction of IFT when the brine is changed from connate water to seawater. Though the IFT changes from 40 to 33 dyne/cm, the reduction changing fluid/fluid interaction is not enough to modify wettability. In the contact angle measurements, contact angles are monitored over a period of 2 days. The contact angle for twice-diluted seawater changes from 90 degrees to 80 degrees. When the twice-diluted seawater is switched to 10-times-diluted seawater, the contact angle between the brine and crude oil changes from 80 degrees to 69 degrees. For 20-times- and 100-times-diluted seawaters, there is no more reduction of contact angle. These results demonstrate the potential of wettability modification toward water-wetness when seawater is diluted by factors of 0.5 and 0.1. These observations lead to the two sets of coreflood tests injecting various diluted seawaters into connate water-saturated cores. The secondary recovery of seawater and tertiary recovery of the various diluted seawaters are deployed into cores. Fig. 1.15 illustrates the oil recovery of the first coreflooding experiment. The seawater injection for secondary recovery produces 67% of OOIP. In tertiary mode, the

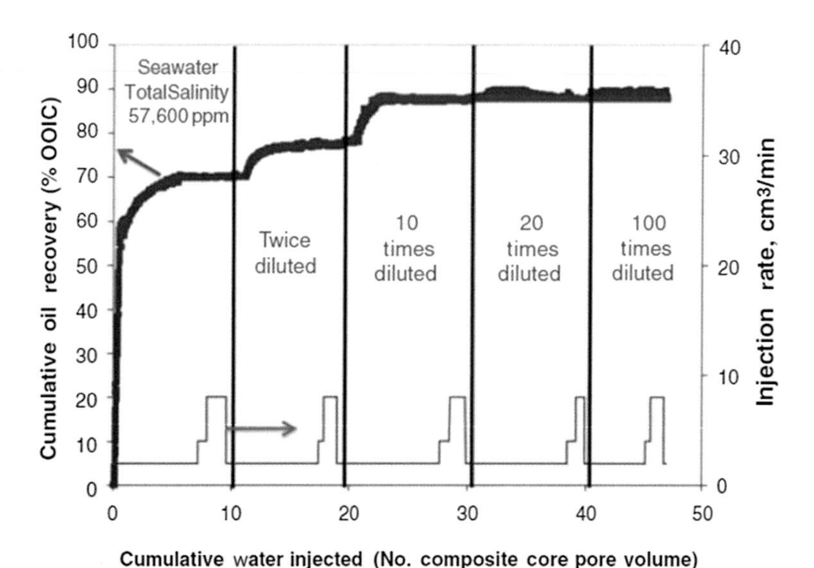

FIG. 1.15 History of oil recovery and injection rate for the first coreflood experiment. (Credit: From Yousef, A. A., Al-Saleh, S. H., Al-Kaabi, A., & Al-Jawfi, M. S. (2011). Laboratory investigation of the impact of injection-water salinity and ionic content on oil recovery from carbonate reservoirs. *SPE Reservoir Evaluation and Engineering, 14*(5), 578–593. https://doi.org/10.2118/137634-PA.)

injections of twice-diluted, 10-times-diluted, and 20-times-diluted seawaters recover the additional oil of 6.99%, 9.12%, and 1.63%, respectively. The last injection of 100-times-diluted seawater provides negligible improvement of oil production. The second coreflooding experiment also reports the increasing oil recovery for tertiary recoveries using twice-diluted, 10-times-diluted, and 20-times-diluted seawaters. The tertiary injection of 100-times-diluted seawater is also ineffective for improving oil recovery. These observations from two sets of coreflooding experiments confirm and validate the potential of various diluted seawater injections for EOR. The NMR experiments measure the distribution of T_2 values, which indicates the pore size distribution in carbonate rocks. The NMR tests investigate the cores, which are saturated with connate water or used in the previous coreflooding experiments. In the results of NMR tests, core saturated with connate water is determined to have macropore and micropore distributions. In the NMR tests using the core from the first coreflooding experiment, it is observed that seawater injection changes the pore distributions in the core. The results are explained that the connectivity between macropore and micropore is enhanced because of the injection of various diluted seawaters. Another NMR experiments measure the distribution of T_2 values of two cores, before and after cleaning process. The negligible change on distribution of T_2 values is observed. This observation clearly indicates that the cleaning process has a negligible effect on T_2 values of NMR. Theses NMR tests prove that the significant change of T_2 values is attributed to the ionic composition of water and salinity and ionic composition of the water affects the carbonate rock surface. This study concluded that injections of various diluted versions of seawater have the potential to impact the rock-fluid interactions and to alter surface charges of carbonate rocks. Therefore, the alteration of the surface charge modifies the wettability of a carbonate rock and enhances oil production from carbonate rocks.

Yousef, Al-Saleh, and Al-Jawfi (2012) validated the previous study of Yousef et al. (2011) and evaluated the potential of the smart waterflood as secondary recovery. In addition, they configured the impact of temperature on the wettability modification during smart waterflood. The carbonate core is determined to be composed of 85% calcite, 12% dolomite, and 3% anhydrite from an XRD analysis. In the previous coreflood test of tertiary mode, the injections of twice-diluted, 10-times-diluted, and 100-times-diluted seawaters recover the additional oil recovery up to 9% over the secondary injection of seawater. In the test of secondary mode, smart waterflood injecting 10-times-diluted water improves the oil recovery by 10% compared with the secondary injection of seawater. These results confirm the potential of smart waterflood for secondary recovery as well as tertiary recovery. This study also investigated the temperature effect on the wettability modification of smart waterflood and described the roles of the temperature. The new experiment using a core, which has low temperature with 135°F, is carried out. The experiments are compared with the high-temperature experiments of Yousef et al. (2011). The experimental temperature of the Yousef et al. (2011) is 212°F. Firstly, ζ potential measurement using diluted seawaters describes that higher temperature condition shifts the surface charge of carbonate rock toward negative, potentially releasing the adsorbed carboxylic components of oil from a carbonate rock surface. Secondly, it is explained that temperature-dependent anhydrite dissolution influences the ionic composition of initial formation brine. The sensitivity of anhydrite dissolution to temperature changes the concentrations of SO_4^{2-} as well as Ca^{2+} in initial formation brine. Generally, anhydrite shows the more dissolution with a decreasing temperature and the anhydrate dissolution produces Ca^{2+} and SO_4^{2-}. The higher concentration of SO_4^{2-} in low temperature condition is confirmed. In terms of the other potential-determining ions, the formation brine in high temperature has the lower concentration of Ca^{2+} and higher concentration of Mg^{2+} compared with the brine in low temperature condition. It is explained that the concentrations of Ca^{2+} and Mg^{2+} are determined by the relative affinities of cations on the rock surface. The activity of Mg^{2+} with rock surface increases in high temperature and the Mg^{2+} substitutes the Ca^{2+} onto the carbonate rock surface. The study concluded that the anhydrite dissolution and the substitution result in the different ionic compositions of the formation brine according to the temperature.

Alotaibi and Yousef (2017) exploited an advanced technology to measure the surface charges of carbonate and crude oil at different ionic composition and temperature conditions. The advanced technique of phase-analysis light scattering (ζ-PALS) could measure the electrophoretic mobility of charged, colloidal suspensions, and oil droplets. In this study, more attention is given to the analysis of how cations and anions, in the equivalent salinity condition, impact on crude oil and carbonate rock, respectively. The study examined the various brines of NaCl, $CaCl_2$, $MgCl_2$, Na_2SO_4, smart water, key ions, and deionized water. Except for brine of deionized water, all brines have the equivalent TDS with 5761 ppm. Brines of key ions and smart water have all ions of Ca^{2+}, Mg^{2+}, and SO_4^{2-} as well as NaCl. The brine of key ions has 3.5 times higher concentration of divalent ions and lower concentration of NaCl than smart water brine. The ζ-PALS estimates the ζ-potential of binary systems of brine/oil droplet and brine/calcite. Generally, the opposite charge of the binary systems results in the electrostatic attraction between the two interfaces of brine/oil droplet and brine/calcite. The attraction makes the oil droplet adheres to rock surface. When both interfaces have negative charge, an electrostatic repulsion occurs and a brine film stabilizes between the interfaces modifying wettability toward water-wetness. The experiments report that the binary system of brine/oil droplet shows the negative ζ-potential regardless of brine type. In the binary system of brine/calcite, the cases of NaCl, Na_2SO_4, and smart water show the negative ζ-potential, but the other show the positive ζ-potential. The increasing positive ζ-potential for key ions brine is attributed to the higher potential of Ca^{2+} above the potential of SO_4^{2-}. This study also explained that the reason of the positive ζ-potential of individual divalent cation brines ($CaCl_2$ and $MgCl_2$) is originated from the adsorption of Ca^{2+} on the calcite surface or the formulation of a surface layer of Mg-bearing calcite. Referring the Huang, FowkesLloyd, and Sanders (1991), adsorption of Ca^{2+} can be explained. Huang et al. (1991) experimentally observed the proportional relationship between adsorption of Ca^{2+} and ζ-potential of calcium carbonate dispersion system (Fig. 1.16). It indicates that the Ca^{2+} could penetrate into hydrolysis layer, substitute the water at calcite surface, and bond at the surface. Ayirala, Al-Saleh, Enezi, and Ali (2018) carried out the additional streaming potential measurements to measure ζ-potential of binary system with various brines and carbonates at different temperatures. It is observed that SO_4^{2-} has a favorable effect on the alteration of ζ-potential toward more negative and its reactivity significantly increases by nearly one order of magnitude in the higher temperature. In addition, the experiments report a few observations regarding roles of cations. The Ca^{2+} has tendency to increase the positive ζ-potential, but the Na^+ and Mg^{2+} slightly change the potential toward less negative. The significant concentration of Na^{2+} has the potential to hinder the reactivity of SO_4^{2-} in the high temperature. The Mg^{2+} and Na^+ have the limited reactivity on carbonate surface regardless of temperature conditions, and the Ca^{2+} is determined to be the most reactive cation making the positive ζ-potential at the carbonate rock surface. The study also evaluated the various brines of NaCl,

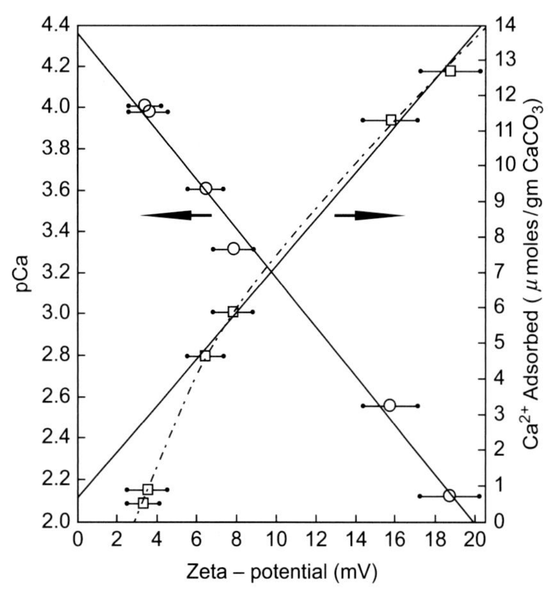

FIG. 1.16 The ζ-potential of the calcium carbonate dispersion system as a function of equilibrium concentration of Ca^{2+} and the adsorption of Ca^{2+}. (Credit: From Huang, Y. C., Fowkes, F. M., Lloyd, T. B., & Sanders, N. D. (1991). Adsorption of calcium ions from calcium chloride solutions onto calcium carbonate particles. *Langmuir*, 7(8):1742–1748. https://doi.org/10.1021/la00056a028.)

$CaCl_2$, $MgCl_2$, Na_2SO_4, key ions, and deionized water, which were tested in Alotaibi and Yousef (2017). The measured ζ-potentials using the brines from the streaming potential measurement are compared with that from ζ-PALS technique. The results from ζ-PALS technique are referred from Alotaibi and Yousef (2017). The streaming potential measurement shows the opposite ζ-potentials of the two brines of $MgCl_2$ and key ions compared with the ζ-PALS technique. Although ζ-PALS technique results in the positive ζ-potentials for the brines, streaming potential measurement shows the negative potentials. The study explained that the negative potentials from the streaming potential measurement agree well with the results of previous coreflood and spontaneous tests (Strand et al., 2006; Zhang & Austad, 2006; Zhang et al., 2007). Therefore, the study concluded that the ζ-PALS technique is comparable to the contact angle measurement and streaming potential measurement is more appropriate to interpret the ζ-potential in coreflood and spontaneous tests.

Based on the experimental observations, LSWF or smart waterflood have been deployed in field tests. Next section illustrates the observations of field trial tests in sandstones and carbonates.

FIELD APPLICATIONS

The previous experiments have observed the increasing oil recovery in the secondary and tertiary modes of LSWF. They have conducted the comprehensive experiments varying the ionic composition of brine, temperature, salinity, initial saturation, and crude oil composition, etc. to explain the oil recovery of LSWF. Based on extensive experimental observations, LSWF has been deployed in fields. This section describes the field applications of LSWF in sandstone and carbonate reservoirs.

Sandstone

Webb, Black, and Al-Ajeel (2003) from British Petroleum (BP) deployed the field test of LSWF in the Middle East sandstone. The study meticulously executed the log-inject-log test and reported the reduction of residual oil saturation by LSWF. The reservoir has quartz contents ranging from 70% to 95%, and kaolinite, plagioclase, illite, and smectite occupy the remainder. It has an average porosity of 0.2 and horizontal permeabilities from 200 to 700 mD with vertical variation because of changes in texture and ductile content. It has an average connate water saturation with 23% and a strong aquifer with some water influx. The salinity of the aquifer water is approximately 250,000 ppm. The log-inject-log test incorporates the pulsed neutron capture (PNC) log because of casing. The PNC log measures the residual oil saturation in near well bore region. The three saline brines of 220,000, 120,000, and 3000 ppm are used for the field test. The test firstly injects high-salinity brine with the 700 bbls, approximately 10–15 PV, to displace the movable oil away from the well bore. For the injections of intermediate-salinity and low-salinity brines, approximately 300 bbls are injected to displace the additional movable oils because of the LSWF effects. After each injection, the response of thermal decay time (TDT) is measured to calculate the residual oil saturation. The test clearly observes that the LSWF decreases remaining oil saturation by up to 50% (Fig. 1.17) and confirms the potential of LSWF in sandstone reservoirs.

McGuire, Chatham, Paskvan, Sommer, and Carini (2005) reported the preliminary corefloods and the four sets of single-well chemical tracer test (SWCTT) in Alaska. The SWCTT is an in situ method to measure fluid saturation, mostly residual oil saturation, in reservoir intervals after waterflood or EOR applications. A brief description of SWCTT follows. The test injects chemical tracer-bearing fluid into the reservoir formation and then produces back the fluid through the same well. The chemical tracer is a reactive partitioning tracer, an ester (typically ethyl acetate). The ester, a primary tracer, is displaced by the large volume of tracer-free water to the target depth, typically 10–15 ft radius, from the wellbore. Subsequent shut-in period allows the ester to dissolve into the water and partly into the oil. Generally, the 10%–50% of ester reacts with water and hydrolyzes into alcohol, a secondary tracer. After the shut-in period, the production of the

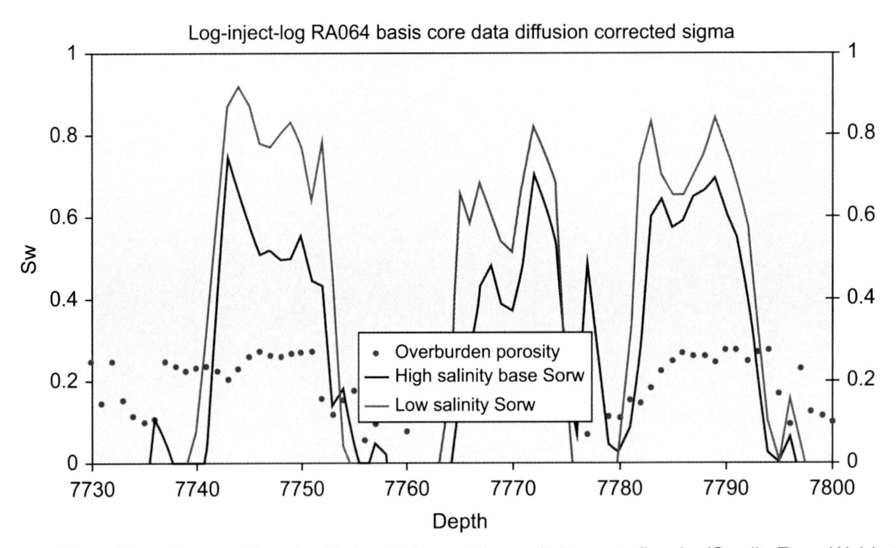

FIG. 1.17 Remaining oil saturations for high-salinity and low-salinity waterfloods. (Credit: From Webb, K. J., Black, C. J. J., & Al-Ajeel, H. (2003). Low salinity oil recovery – log-inject-log. Paper presented at the Middle East Oil show, Bahrain, 9–12 June. https://doi.org/10.2118/81460-MS.)

well initiates and gas chromatography measures the unreacted ester, ethanol, and propanol content. During the production, the remaining ester in reservoir is delayed because of the partition between the immobile residual oil and the mobile water. However, the secondary tracer directly flows back to the well without a delay. The delay of production of the ester is related to the residual oil saturation. Various interpretation methods with the separation between the ester and secondary tracer productions estimate the residual oil saturation. McGuire et al. (2005) described the two sets of SWCTTs in the Ivishak sandstone, and each set in the Kuparuk and Kekiktuk sandstones, respectively. In Prudhoe Bay Unit, Kuparuck reservoir in L-122 has the reservoir temperature of 150°C and the average porosity of 16%. A set of two SWCTTs measures the residual oil saturations after high-salinity waterflood with 23,000 ppm and LSWF with 3000 ppm, respectively. According to the results of SWCTTs (Fig. 1.18), the high-salinity waterflood and LSWF result in the residual oil saturations of approximately 0.21 and 0.13, respectively. In the Northwest Eileen of the Prudhoe Bay Field, the Ivishak Zone 4 sand formation has a temperature of 217°F. The two SWCTTs in the reservoir show that high-salinity waterflood with 23,000 ppm approximately remains the residual oil saturation of 0.19 and LSWF with 3000 ppm reduces the residual oil saturation by 0.04. The another Kekiktuk reservoir in the Endicott Field has 210°F and the average porosity of 0.24. Two SWCTTs estimate the residual oil saturations of about 0.43 after high-salinity waterflood and about 0.34 after LSWF. The last set deploys the three SWCTTs in Ivishak Zone 4B formation in the Prudhoe Bay Field. The test observes the residual oil saturations after injections of high-salinity, intermediate-salinity, and low-salinity brines. High-salinity waterflood with 22,000 ppm approximately results in the residual oil saturation of 0.21. Intermediate-salinity waterflood with 7000 ppm hardly shows any discernible reduction in the residual oil saturation. However, the LSWF recovers additional oil and reduces the residual oil saturation of 0.04. This study clearly concluded that LSWF has the potential modifying wettability and reducing residual oil saturation, and there is an effective salinity level of 5000 ppm TDS or less to observe the potential in sandstone reservoirs.

The Lager, Webb, Collins, et al. (2008) from BP deployed the LSWF at interwell test as well as SWCTT in an Alaskan oil field and confirmed the direct field evidences for EOR potential. A single hydraulic unit in the field is subject to the interwell test of LSWF. An injector (MPL-16A) and two nearby producers (MPL-07 and MPL-11) in the targeted hydraulic unit are candidate wells and monitored in the test. The unit has natural depletion production for 4 years. A following waterflood, injecting brackish brine with 16,640 ppm TDS, increases the oil production of the producer (MPL-07) from 400 to 1100 bbls/day. After 3 years of the waterflood, oil production decreases and water-cut increases up to 95%. An additional injection of miscible additive increases oil production from 200 to 500 bbls/day for a while, but the production rate falls off to 150 bbls/day after a year and a half. To increase the oil production rate and decrease water-cut, the LSWF with 2600 ppm TDS is performed in the hydraulic unit. The oil production reaches to the maximum rate of 320 bbls/day and decreases to 200 bbls/day (Fig. 1.19). Water-cut also drops from 92% to 87%. In addition, the injection of low-salinity water shows the constant injectivity indicating no formation damage due to clay swelling or fines generation. In the analysis of produced water from the MPL 07 producer, the Mg^{2+} is completely removed in the effluent water and the removal is regarded as the proof of interaction between the reservoir rock and injecting brine. Based on the field observations, a new well is drilled to conduct the SWCTT. The tests are performed injecting the four different brines: high-salinity water, produced water from MPL-07, Prine Creek aquifer water, and optimized low-salinity water in a sequence of decreasing salinity. The waterfloods using both high salinity and MPL-07 produced water, approximately, remain the residual oil saturation of 0.3. The injection of aquifer water makes the residual oil saturation of 0.2. The injection of optimized low-salinity water lowers the residual oil saturation by 0.08. The both field tests obviously confirm the EOR of LSWF and are in line with the previous experimental observations.

The BP has performed more field tests of LSWF. Seccombe, Lager, Webb, Jerauld, and Fueg (2008) described the five sets of SWCTTs at Endicott Field. Because the LSWF is affected by reservoir mineralogy, the study examined the Endicott petrology by SEM and wireline log. The SEM photomicrograph describes that the Kekituk formation at Endicott Field is composed of seven pore-filling constituents. Primary constituent is quartz and secondary is kaolinite clay. The remainder less than 1% is attributed to calcite, dolomite, siderite, pyrite, and argillaceous matrix. The wireline log test assumes that the formation has only pure quartz and kaolinite clay, neglecting the remaining constituents. The test estimates the content of kaolinite clay in the test zone. This study correlated the concentration of kaolinite clay with the enhanced oil recovery

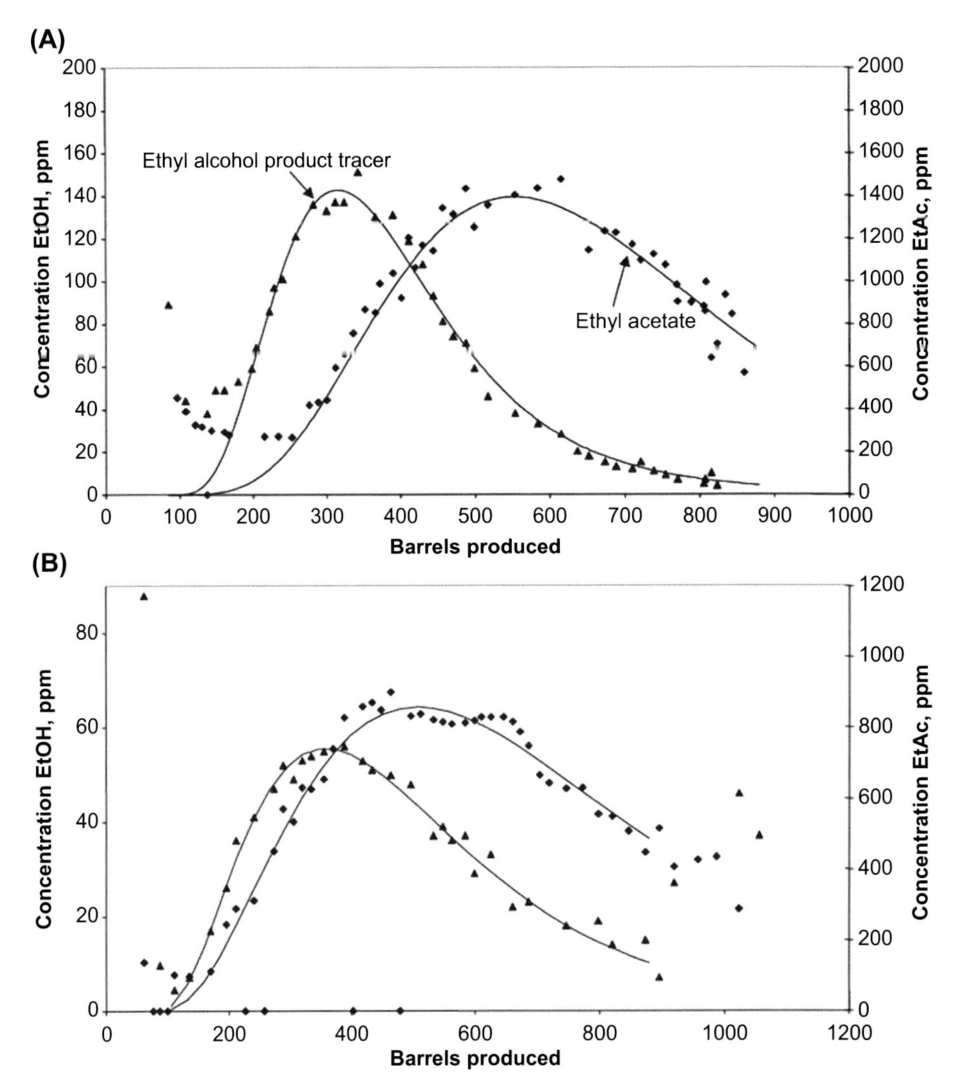

FIG. 1.18 Results of two SWCTTs of **(A)** high-salinity waterflood and **(B)** low-salinity waterflood. (Credit: From McGuire, P. L., Chatham, J. R., Paskvan, F. K., Sommer, D. M., & Carini, F. H. (2005). Low salinity oil recovery: An exciting new EOR opportunity for Alaska's North slope. *Paper presented at the SPE Western Regional Meeting, Irvine, California, USA, 30 March–1 April.* https://doi.org/10.2118/93903-MS.)

of LSWF, observed from the SWCTTs. The SWCTTs for the high-salinity waterflood report that the residual oil saturation ranges from 39% to 43%. The tests are performed to analyze the efficiency of LSWF slug size by varying the slug size of low-salinity water. In the tests, the 2 PV injection of LSWF recovers the additional oil saturation of 0.090.17. Based on the observations, linear relationship between the additional oil recovery by LSWF and the kaolinite concentration is constructed. In addition, a slug size of 0.4 PV of LSWF is determined to be fully effective compared with the case of 2.0 PV in terms of additional oil recovery. The study concluded that the LSWF is an attractive and effective EOR at Endicott Field and the content of kaolinite clay is of importance to activate EOR effect.

Seccombe et al. (2010) reported a few field trials of interwell in the same field. An injector and a producer are candidates for the interwell test. The well pair in the subzone 3A2, Endicott Field, has the interwell distance of 1040 ft. The subzone has relatively higher

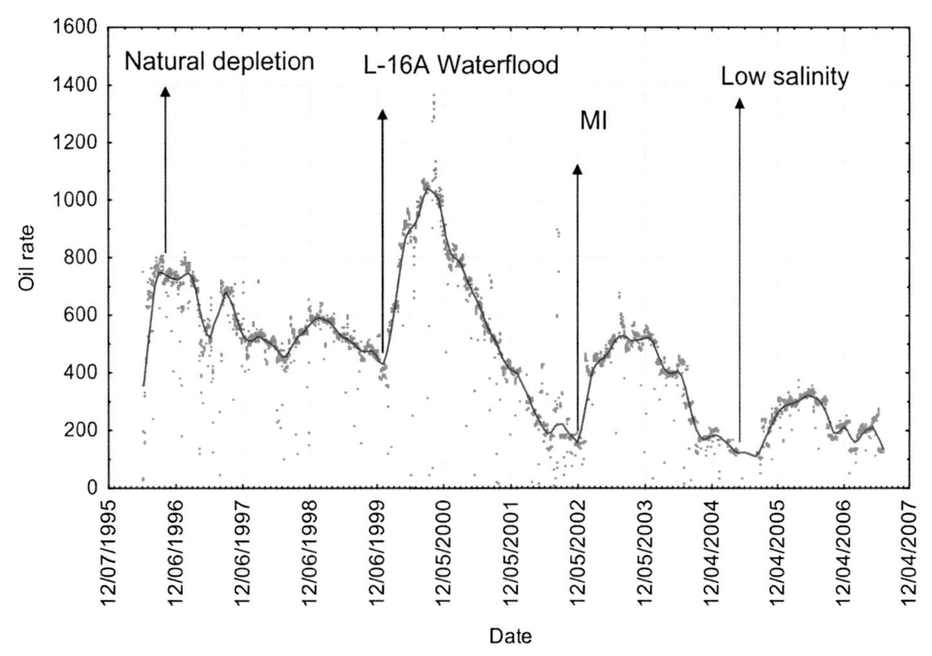

FIG. 1.19 History of oil production rate from a hydraulic unit in the Alaskan oil field. (Credit: From Lager, A., Webb, K. J., Collins, I. R., & Richmond, D. M. (2008b). LoSal enhanced oil recovery: Evidence of enhanced oil recovery at the reservoir scale. *Paper presented at the SPE Symposium on improved oil recovery, Tulsa, Oklahoma, USA*, 20–23 April. https://doi.org/10.2118/113976-MS.)

content of clay and 30–45 ft thickness. Firstly, the high-salinity water using produced water is injected into the reservoir and water-cut reaches to 95% after about 6 months. The LSWF commences and total 1.5 PV of low-salinity water is injected for the next 4 months. Then, high-salinity waterflood resumes as postflush. The oil rate, water-cut, and ionic composition are monitored at the producer. The LSWF increases the oil rate and decreases water-cut with a detection reducing salinity (Fig. 1.20). The water-cut roughly drops from 95% to 92%. The incremental oil production is the 0.1 PV of swept zone. These field observations correspond to the experimental observations and the SWCTT results of Seccombe et al. (2008). The analysis of water chemistry shows that the increasing production of Fe is obtained without any injection of Fe. In addition, the production of Fe increases when the LSWF effect appears. The study related the geochemical role of Fe to wettability. It explained that the Fe bridges crude oil and kaolinite clay mineral and the LSWF disturbs the bridge liberating the crude oil from the clay. An additional analysis interpreting tracer data and water chemistry determines the reduction of residual oil saturation by 0.13 for tertiary LSWF.

Skrettingland, Holt, Tweheyo, and Skjevrak (2011) reported the unsuccessful application of LSWF at the Snorre oil field through experiments and SWCTT. The Snorre oil field has the two Lunde and Statfjord formations of the late Triassic and early Jurassic. The sandstone reservoir has the temperature of 90°C and total clay content in the range of 5%–35%. The XRD experiments indicate that the primary composition of minerals is quartz and the remainders are mostly K-feldspar, plagioclase, and kaolinite. The formations are classified as neutral-wet to weakly water-wet. The oil producer well P-07 in Statfjord formation is subject to the SWCTT. The synthetic seawater of 34,300 ppm TDS and low-salinity water of 440 ppm TDS are injected through the well. The low-salinity water is made by adding 1% seawater in freshwater. The SWCTTs after seawater injection and LSWF approximately result in the residual oil saturations of 0.23. There was no reduction of residual oil saturation by LSWF compared with the seawater injection. These results agree to the laboratory results of LSWF. This study concluded that wetting condition in the Snorre oil field is sufficiently efficient for seawater injection and the initial wetting condition is a crucial factor for the successful LSWF project.

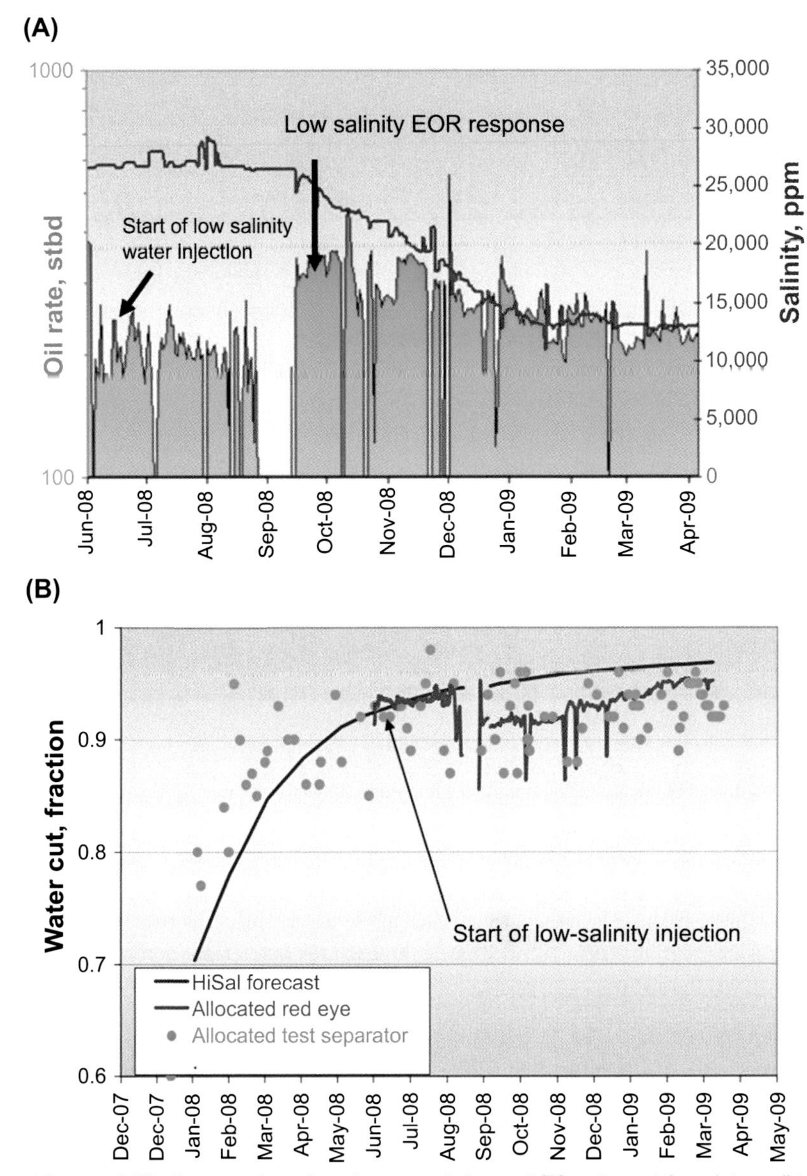

FIG. 1.20 History of **(A)** oil rate and produced water salinity, and **(B)** water-cut from interwell test. (Credit: From Seccombe, J., Lager, A., Gary, J., Jhaveri, B., Todd, B., Bassler, S., et al. (2010). Demonstration of low-salinity EOR at Interwell scale, Endicott field, Alaska. Paper presented at the SPE improved oil recovery Symposium, Tulsa, Oklahoma, USA, 24–28 April. https://doi.org/10.2118/129692-MS.)

Callegaro et al. (2014) reported the SWCTT for LSWF investigation at on-shore field in West Africa. During SWCTTs measuring the residual oil saturations after seawater injection and LSWF, the negligible change of residual oil saturation is observed. The study concluded the minor effect of LSWF is attributed to low concentration of clay and already low residual oil saturation after seawater injection. Rotondi, Callegaro, Franco, and Bartosek (2014) described the another field trials of LSWF for EOR at on-shore field in West Africa.

The target reservoir is the sandstone formation and has the temperature with 88°C. In the first trial of LSWF, the log-inject-log field test and SWCTT are performed based on the experimental and numerical simulation works. Referring the experimental and simulation results, a further technical-economical evaluation provides the salinity level for LSWF. Compared with the case of the seawater injection, the SWCTT reports the negligible reducing residual oil saturation by LSWF. Another giant on-shore brownfield in North Africa is also tested for LSWF efficiency. The SWCTT is implemented to quantify the LSWF potential in the field. Callegaro et al. (2015) reported the detailed results of the field trial in the brownfield. The reservoir has temperature in the ranges of 76−83°C and viscous oil in the ranges of 3−8 cp at the reservoir condition. The original formation salinity is higher than 220 g/L. The primary depletion initiates in 1955, and the secondary seawater injection of 39 g/L salinity is implemented in 1985. When the secondary injection of seawater becomes ineffective with the high water-cut and moderate recovery, it is initiated to evaluate the prospect of the another IOR/EOR technology implementation. The LSWF is evaluated for the tertiary recovery in the field. The positive experimental results of LSWF coreflooding lead to the SWCTT implementation of LSWF in the field. The field tests of SWCTT measure the residual oil saturation after seawater injection and LSWF. Interpretation on the SWCTT tracer data shows that the LSWF reduces residual oil saturation of 5%−11%, compared with seawater injection.

The Burgan field in Kuwait is the second largest in the world and the largest clastic reservoir. It is mainly composed of five giant reservoirs, Wara sandstone, Burgan sandstone, Mauddud limestone, Minagish Oolite limestone, and Marrat carbonate. The Burgan sandstone has the subsections of 3SU, 3SM, 3SL, and 4S. The 3SM and 4S have good quality with up to 10,000 md. The 3SU and 3SL sections of Burgan sandstone and Wara sandstone are more heterogeneous. Kuwait Oil Company (KOC) deployed the field trial of LSWF in the Burgan field. The second largest reservoir has been produced since 1946. Because the field has the strong natural aquifer below and costs little for drilling, the production is most economic until 1988. At that time, water production is observed and additional facilities are introduced with artificial lift. The water production reaches the 23% of liquid production in 2012. To extend the life of Burgan field, KOC reviews the early application of EOR technology appropriate to the field. The early application potentially expects cost cuttings in terms of water handling, disposal, and injection. Abdulla et al. (2013) and Al-Qattan et al. (2018) have reported the field tests of LSWF as EOR application in the Burgan field. Abdulla et al. (2013) described a first SWCTT application of LSWF in the Burgan field. The sole compatibility test between freshwater injection and core is carried out as a preliminary test. The target reservoir is the rock type 1, which is the clean and best quality rock with the least clay. In terms of wettability modification effects by LSWF, the study mainly investigated the residual oil saturation changes without considering relative permeability curve changes. Two wells, BG-A1 in South of the Wara and BG-A2 in the South-West of the 3SU, are candidate wells for SWCTT. The produced water and low-salinity water are injected into each well, and SWCTT is carried out following the injections. The test results of BG-A2 have a good quality, but the results of BG-A1 have some errors to measure the tracers from SWCTT. The tracer data sets of the SWCTTs are interpreted with five methods: (1) three-layer tracer flow simulation with flow irreversibility; (2) one-layer tracer flow simulation with cross-flow; (3) simple calculations from peak arrival times/volumes; (4) direct shift; and (5) full reservoir simulation. The two methods of simple calculations from peak arrival times/volumes and reservoir simulation show unclear interpretations because of an unclear peak of tracer data and the complex input parameters of reservoir simulation. In the SWCTTs of BG-A1, the three-layer tracer flow simulation calculates the residual oil saturation of about 0.12, which is higher than the residual oil saturations calculated by other methods. The SWCTTs from BG-A2 provide relatively consistent residual oil saturation regardless of interpretation methods. This study concluded that the applications of LSWF in BG-A1 and BG-A2 approximately reduce the average residual oil saturation by 0.3 in the clean and good reservoir in Burgan field. As reported by Al-Qattan et al. (2018), these results lead to the implementation of additional SWCTT of LSWF in rock type 2 of the Wara formation. The rock type 2 is characterized with higher clay content expecting the more potential of LSWF. The test well, Well A, shows the oil cut of approximately 66% before SWCTT trial. The PV of target depth of SWCTT is estimated with 500 bbls. The target interval of 16 ft has an average porosity of 0.23, and the radius of investigation is about 15.6 ft. In the well, the high-salinity water injection of up to 148,000 ppm TDS displaces all movable oil, before

SWCTT application. The first SWCTT follows the high-salinity water injection and provides a baseline of the residual oil saturation to compare the residual oil saturation after LSWF. The second SWCTT aims to monitor the reduction of residual oil saturation by LSWF. The LSWF with 2 PV is applied into the Well A, before the second SWCTT. Using the three-layer tracer flow simulation method, the tracer data sets from the two SWCTTs indicate that LSWF reduces residual oil saturation by 0.03.

Previous field trials including log-inject-log and SWCTT demonstrate that LSWF sufficiently has potential to reduce residual oil saturation when the sandstone reservoir satisfies some perquisites.

Carbonate Rocks

Compared with the previous studies of sandstone, to the best of our knowledge, only one study published the field trials of LSWF in carbonate reservoirs. Yousef et al. (2012) reported the first field test of LSWF, i.e., SmartWater, in a carbonate reservoir based on the experimental observations (Yousef, Al-Saleh, and Al-Jawfi 2012; Yousef et al. 2011). The previous experimental observations concluded that injections of the twice-diluted seawater and 10-times-diluted seawater have EOR potential. Yousef et al. (2012) validated the two types of diluted seawater as the candidate brines of LSWF through SWCTT. The study reported a preliminary study of numerical simulation and a real field implementation. It designed to deploy the SWCTT and LSWF in two potential wells, Well A and Well B. Different scenarios of SWCTT were designed for each well. The test in Well A has a plan with three trials of SWCTT to observe residual oil saturation reduction by LSWF with 10-times-diluted seawater. The first trial of SWCTT measures the residual oil saturation after seawater injection. The second trial confirms the results of the first SWCTT. The last trial measures the saturation after 10-times-diluted seawater injection. The test in Well B is also designed with three trials of SWCTT. The first trial detects the residual oil saturation after seawater injection. The second one captures residual oil saturation reduction by the twice-diluted seawater injection. The third trial has an objective to determine the residual oil saturation after LSWF with 10-times-diluted seawater. Firstly, a preliminary study numerically simulates and predicts the two different scenarios of SWCTTs in Well A and Well B using experimental measurements, before real field implementations. The preliminary study interprets the tracer data obtained from the numerical simulations of SWCTTs and estimates the residual oil saturation reductions by LSWF using an analytical method and a direct determination method. In the Well A, LSWF with twice-diluted seawater is numerically determined to reduce residual oil saturation by 0.06–0.07. In the Well B, numerical models and both interpretation methods of SWCTT predict the reduction up to 0.03 for twice-diluted seawater injection. For the injection of 10-times-diluted seawater, the analytical method predicts the additional reduction of residual oil saturation by 0.04, but the direct method calculates the reduction by 0.06. Considering the uncertainty of the SWCTT, the numerical simulations of LSWF in the Well B predict the total reduction of 0.07–0.09. With the clear confirmation of numerical simulations, the real field trials of SWCTT are executed in Well A and Well B. In the real field trials, the residual oil saturations are calculated by a history match method and the analytical method. In the Well A, the tracer results of SWCTT show good quality. Both interpretation methods of the analytical and the history match methods using the real tracer results, consistently, show that LSWF of twice-diluted seawater reduces the residual oil saturation by 0.07, compared with seawater injection (Fig. 1.21). In addition, the real field test and the preliminary study of numerical simulation show the consistent results in the Well A. In the Well B, a field implementation of LSWF with twice-diluted seawater lowers the residual oil saturation by 0.03 in the test zone and that with 10-times-diluted seawater additionally decreases the residual oil saturation by 0.03. In comparison with the preliminary numerical simulation, the real field test shows a different reduction of residual oil saturation for the injection of 10-times-diluted seawater. This difference is within the range of the uncertainty of the implementation of SWCTT test. This study concluded and demonstrated that the field trials of SWCTT agree with experimental observations and reducing residual oil saturation in carbonate reservoirs can be achieved by diluted seawater injection.

The extensive LSWF research studies have reported various experimental evidences for IOR/EOR and validated the potential to reduce residual oil saturation through the field tests. The research studies have formulated the potential mechanisms of LSWF to explain the IOR/EOR observations in sandstone and carbonate reservoirs. Hence, next chapter discusses the up-to-date mechanisms proposed in sandstone and carbonate reservoirs.

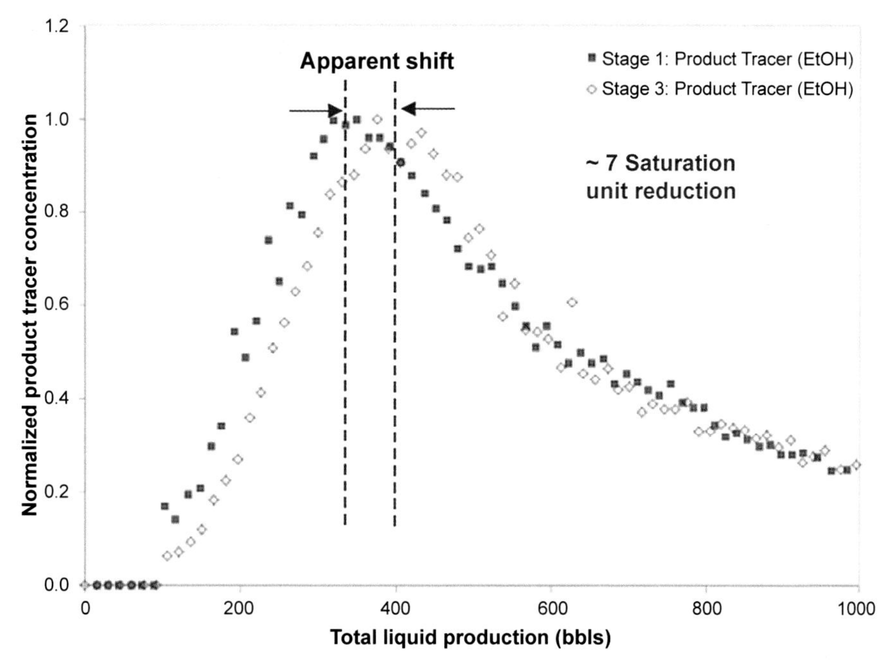

FIG. 1.21 Comparison of field-measured tracer concentration profiles of SWCTT between seawater and 10-times-diluted seawater injections for Well A. (Credit: From Yousef, A. A., Liu, J. S., Blanchard, G. W., Al-Saleh, S., Al-Zahrani, T., Al-Zahrani, R. M., et al. (2012b). Smart waterflooding: Industry. Paper presented at the SPE Annual technical conference and exhibition, San Antonio, Texas, USA, 8–10 Ocotober. https://doi.org/10.2118/159526-MS.)

REFERENCES

Abdulla, F., Hashem, H. S., Abdulraheem, B., Al-Nnaqi, M., Al-Qattan, A., John, H., et al. (2013). First EOR trial using low salinity water injection in the greater Burgan field, Kuwait. In *Paper presented at the SPE Middle East oil and gas show and conference, Manama, Bahrain, 10–13 March.* https://doi.org/10.2118/164341-MS.

Agbalaka, C. C., Dandekar, A. Y., Patil, S. L., Khataniar, S., & Hemsath, J. R. (2008). Coreflooding studies to evaluate the impact of salinity and wettability on oil recovery efficiency. *Transport in Porous Media, 76*, 77–94. https://doi.org/10.1007/s11242-008-9235-7.

Al-Qattan, A., Sanaseeri, A., Al-Saleh, Z., Singh, B. B., Al-Kaaoud, H., Delshad, M., et al. (2018). Low salinity waterflood and low salinity polymer injection in the Wara reservoir of the greater Burgan field. In *Paper presented at the SPE EOR conference at oil and gas West Asia, Muscat, Oman, 26–28 March.* https://doi.org/10.2118/190481-MS.

Alotaibi, M. B., & Yousef, A. (2017). The role of individual and combined ions in waterflooding carbonate reservoirs: Electrokinetic study. *SPE Reservoir Evaluation and Engineering, 20*(1), 77–86. https://doi.org/10.2118/177983-PA.

Austad, T., Rezaeidoust, A., & Puntervold, T. (2010). Chemical mechanism of low salinity water flooding in sandstone reservoirs. In *Paper presented at the SPE improved oil recovery symposium, Tulsa, Oklahoma, USA, 24–28 April.* https://doi.org/10.2118/129767-MS.

Austad, T., Strand, S., Høgnesen, E. J., & Zhang, P. (2005). Seawater as IOR fluid in fractured chalk. In *Paper presented at the SPE International symposium on oilfield chemistry, The Woodlands, Texas, USA, 2–4 February.* https://doi.org/10.2118/93000-MS.

Ayirala, S., Al-Saleh, S. H., Enezi, S., & Ali, Y. (2018). Effect of salinity and water ions on electrokinetic interactions in carbonate reservoir cores at elevated temperatures. *SPE Reservoir Evaluation and Engineering, 21*(3), 114. https://doi.org/10.2118/189444-PA.

Berg, S., Cense, A. W., Jansen, E., & Bakker, K. (2010). Direct experimental evidence of wettability modification by low salinity. *Petrophysics, 51*(5), 314–322.

Bernard, G. G. (1967). Effect of floodwater salinity on recovery of oil from cores containing clays. In *Paper presented at the SPE California Regional Meeting, Los Angeles, California, USA, 26–27 October.* https://doi.org/10.2118/1725-MS.

Callegaro, C., Bartosek, M., Nobili, M., Franco, M., Pollero, M., Baz, D. M. M., et al. (2015). Design and implementation of low salinity waterflood in a North African Brown field. In *Paper presented at the Abu Dhabi International petroleum exhibition and conference, Abu Dhabi, UAE, 9–12 November.* https://doi.org/10.2118/177590-MS.

Callegaro, C., Franco, M., Bartosek, M., Buscaglia, R., Visintin, R., Hartvig, S. K., et al. (2014). Single well chemical tracer tests to assess low salinity water and surfactant EOR processes in West Africa. In *Paper presented at the International petroleum technology conference, Kuala Lumpur, Malaysia, 10–12 December.* https://doi.org/10.2523/IPTC-17951-MS.

Hallenbeck, L. D., Sylte, J. E., Ebbs, D. J., & Thomas, L. K. (1991). Implementation of the Ekofisk field waterflood. *SPE Formation Evaluation, 6*(3), 284–290. https://doi.org/10.2118/19838-PA.

Huang, Y. C., Fowkes, F. M., Lloyd, T. B., & Sanders, N. D. (1991). Adsorption of calcium ions from calcium chloride solutions onto calcium carbonate particles. *Langmuir, 7*(8), 1742–1748. https://doi.org/10.1021/la00056a028.

Jadhunandan, P. P., & Morrow, N. R. (1995). Effect of wettability on waterflood recovery for crude-oil/brine/rock systems. *SPE Reservoir Engineering, 10*(1), 40–46. https://doi.org/10.2118/22597-PA.

Lager, A., Webb, K. J., Black, C. J. J., Singleton, M., & Sorbie, K. S. (2008a). Low salinity oil recovery – an experimental investigation1. *Petrophysics, 49*(1), 28–38. https://doi.org/10.2118/93903-MS.

Lager, A., Webb, K. J., Collins, I. R., & Richmond, D. M. (2008b). LoSal enhanced oil recovery: Evidence of enhanced oil recovery at the reservoir scale. In *Paper presented at the SPE symposium on improved oil recovery, Tulsa, Oklahoma, USA, 20–23 April.* https://doi.org/10.2118/113976-MS.

Ligthelm, D. J., Gronsveld, J., Hofman, J., Brussee, N., Marcelis, F., & van der Linde, H. (2009). Novel waterflooding strategy by manipulation of injection brine composition. In *Paper presented at the EUROPEC/EAGE conference and exhibition, Amsterdam, Netherlands, 8–11 June.* https://doi.org/10.2118/119835-MS.

McGuire, P. L., Chatham, J. R., Paskvan, F. K., Sommer, D. M., & Carini, F. H. (2005). Low salinity oil recovery: An exciting new EOR opportunity for Alaska's North slope. In *Paper presented at the SPE Western Regional Meeting, Irvine, California, USA, 30 March–1 April.* https://doi.org/10.2118/93903-MS.

Morrow, N. R. (1990). Wettability and its effect on oil recovery. *Journal of Petroleum Technology, 42*(12), 1476–1484. https://doi.org/10.2118/21621-PA.

Nasralla, R. A., & Nasr-El-Din, H. A. (2014). Double-layer expansion: Is it a primary mechanism of improved oil recovery by low-salinity waterflooding? *SPE Reservoir Evaluation and Engineering, 17*(01), 49–59. https://doi.org/10.2118/154334-PA.

RezaeiDoust, A., Puntervold, T., & Austad, T. (2011). Chemical verification of the EOR mechanism by using low saline/smart water in sandstone. *Energy and Fuels, 25*(5), 2151–2162. https://doi.org/10.1021/ef200215y.

Rotondi, M., Callegaro, C., Franco, M., & Bartosek, M. (2014). Low salinity water injection: Eni's experience. In *Paper presented at the Abu Dhabi international petroleum exhibition and conference, Abu Dhabi, UAE, 10–13 November.* https://doi.org/10.2118/171794-MS.

Seccombe, J., Lager, A., Gary, J., Jhaveri, B., Todd, B., Bassler, S., et al. (2010). Demonstration of low-salinity EOR at interwell scale, Endicott field, Alaska. In *Paper presented at the SPE improved oil recovery symposium, Tulsa, Oklahoma, USA, 24–28 April.* https://doi.org/10.2118/129692-MS.

Seccombe, J. C., Lager, A., Webb, K. J., Jerauld, G., & Fueg, E. (2008). Improving waterflood recovery: LoSalTM EOR field evaluation. In *Paper presented at the SPE symposium on improved oil recovery, Tulsa, Oklahoma, USA, 20–23 April.* https://doi.org/10.2118/113480-MS.

Shehata, A. M., & Nasr-El-Din, H. A. (2017). Laboratory investigations to determine the effect of connate-water composition on low-salinity waterflooding in sandstone reservoirs. *SPE Reservoir Evaluation and Engineering, 20*(01), 59–76. https://doi.org/10.2118/171690-PA.

Skrettingland, K., Holt, T., Tweheyo, M. T., & Skjevrak, I. (2011). Snorre low-salinity-water injection–coreflooding experiments and single-well field pilot. *SPE Reservoir Evaluation and Engineering, 14*(02), 182–192. https://doi.org/10.2118/129877-PA.

Strand, S., Høgnesen, E. J., & Austad, T. (2006). Wettability alteration of carbonates—effects of potential determining ions (Ca^{2+} and SO_4^{2-}) and temperature. *Colloids and Surfaces A: Physicochemical and Engineering Aspects, 275*(1), 110. https://doi.org/10.1016/j.colsurfa.2005.10.061.

Tang, G. Q., & Morrow, N. R. (1997). Salinity, temperature, oil composition, and oil recovery by waterflooding. *SPE Reservoir Engineering, 12*(04), 269–276. https://doi.org/10.2118/36680-PA.

Tang, G.-Q., & Morrow, N. R. (1999). Influence of brine composition and fines migration on crude oil/brine/rock interactions and oil recovery. *Journal of Petroleum Science and Engineering, 24*(2), 99–111. https://doi.org/10.1016/S0920-4105(99)00034-0.

Webb, K. J., Black, C. J. J., & Al-Ajeel, H. (2003). Low salinity oil recovery - log-inject-log. In *Paper presented at the Middle East oil show, Bahrain, 9–12 June.* https://doi.org/10.2118/81460-MS.

Yildiz, H. O., & Morrow, N. R. (1996). Effect of brine composition on recovery of Moutray crude oil by waterflooding. *Journal of Petroleum Science and Engineering, 14*(3), 159–168. https://doi.org/10.1016/0920-4105(95)00041-0.

Yousef, A. A., Al-Saleh, S., & Al-Jawfi, M. S. (2012a). Improved/enhanced oil recovery from carbonate reservoirs by tuning injection water salinity and ionic content. In *Paper presented at the SPE improved oil recovery symposium, Tulsa, Oklahoma, USA, 14–18 April.* https://doi.org/10.2118/154076-MS.

Yousef, A. A., Al-Saleh, S. H., Al-Kaabi, A., & Al-Jawfi, M. S. (2011). Laboratory investigation of the impact of injection-water salinity and ionic content on oil recovery from carbonate reservoirs. *SPE Reservoir Evaluation and Engineering, 14*(5), 578–593. https://doi.org/10.2118/137634-PA.

Yousef, A. A., Liu, J. S., Blanchard, G. W., Al-Saleh, S., Al-Zahrani, T., Al-Zahrani, R. M., et al. (2012b). Smart waterflooding: Industry. In *Paper presented at the SPE annual technical conference and exhibition, San Antonio, Texas, USA, 8–10 Ocotober*. https://doi.org/10.2118/159526-MS.

Zhang, P., & Austad, T. (2006). Wettability and oil recovery from carbonates: Effects of temperature and potential determining ions. *Colloids and Surfaces A: Physicochemical and Engineering Aspects, 279*(1), 179–187. https://doi.org/10.1016/j.colsurfa.2006.01.009.

Zhang, P., Tweheyo, M. T., & Austad, T. (2007). Wettability alteration and improved oil recovery by spontaneous imbibition of seawater into chalk: Impact of the potential determining ions Ca^{2+}, Mg^{2+}, and SO_4^{2-}. *Colloids and Surfaces A: Physicochemical and Engineering Aspects, 301*(1), 199–208. https://doi.org/10.1016/j.colsurfa.2006.12.058.

CHAPTER 2

Mechanisms of Low-Salinity and Smart Waterflood

ABSTRACT

The enhanced oil production of low-salinity waterflood (LSWF) and smart waterflood has been demonstrated through experiments and field-scaled tests for sandstone and carbonate reservoirs. Extensive studies have tried to reveal the exact mechanism underlying the LSWF to increase the oil recovery. A number of theories have been proposed to describe the increasing oil production by LSWF. However, there is a controversy on the exact mechanism of the LSWF. Because of the inherent difference between sandstone and carbonate reservoirs, there is no universal mechanism to describe the LSWF process for both sandstone and carbonate reservoirs. Therefore, this chapter discusses the reliable mechanisms proposed in both sandstone and carbonate reservoirs.

Up to now, numerous mechanisms of low-salinity waterflood (LSWF) and smart waterflood have been proposed. Many studies have reported that wettability modification as the dominant mechanism of LSWF for sandstone and carbonate reservoirs. There are other discussions of mechanism such as the clay particle-plugging high permeable zone in sandstone reservoirs. Because the complex COBR interactions are involved in the process, the effects of the LSWF and smart waterflood can be probably combined results of several different mechanisms contributing together. Extensive studies have formulated a variety of theories to explain the mechanism of LSWF based on the observations of experiments and field test. No single theory has been widely accepted. Some of most promising theories for LSWF and smart waterflood in sandstone and carbonate reservoirs are explained further in this chapter.

MECHANISMS IN SANDSTONE RESERVOIRS

In spite of the extensive research studies of LSWF conducted over two decades, unanimous mechanism has not been accepted to explain the experimental observations in sandstone. The proposed mechanisms include fines migration, in situ generation of surfactant, salting-in, multicomponent ionic exchange, electrical double layer expansion, and pH increase. A brief description of these mechanisms follows.

Fines Migration

Tang and Morrow (1999) proposed the fines migration mechanism based on the experimental observations of sandstone and the mechanism is schematically described in Fig. 2.1. The mechanism hypothesizes that the heavy polar components of crude oil adhere to the clay, which coats the pore walls of sandstone rock grain (Fig. 2.1A). The crude oil has two potential behaviors in waterflooding process. Firstly, crude oil drops adhere to fines at pore walls and they remain as trapped oil fraction. Secondly, the mixed-wet clay particles adhering crude oil are stripped away from the pore walls with the flowing oil and the clay particles tend to be at the oil-water interface (Fig. 2.1B). It is explained that the behavior and stability of mixed-wet fines depend on the balance between mechanical and colloidal forces. The mechanical forces are capillary forces adhering crude oil to the fines and viscous forces stripping clay from the pore wall. In addition, a mechanical resistance mitigates the stripping. The colloidal forces between particles govern the stability of colloids and also control the displacement of oil by changing electrical double layer. Reduction in salinity expands the electrical double layer in the aqueous phase between particles and promotes the stripping of clay. In addition, the mixed-wet clay at the oil-water interface inhibits the residual oil trapping by snap-off. Therefore, the partial removal of mixed-wet clay particles from the pore wall potentially causes the locally heterogeneous wetting and increases the oil recovery (Fig. 2.1C).

In Situ Generation of Surfactant

McGuire, Chatham, Paskvan, Sommer, and Carini (2005) published the results of single-well chemical tracer test (SWCTT) for Alaska's North Slope. The study suggested the in situ generation of surfactant by pH

Hybrid Enhanced Oil Recovery using Smart Waterflooding. https://doi.org/10.1016/B978-0-12-816776-2.00002-7

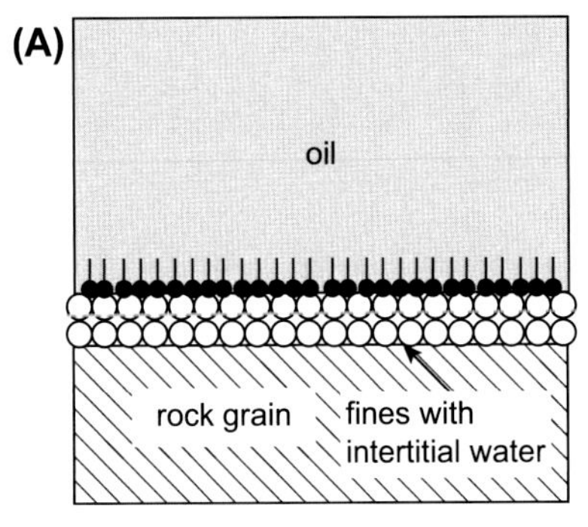

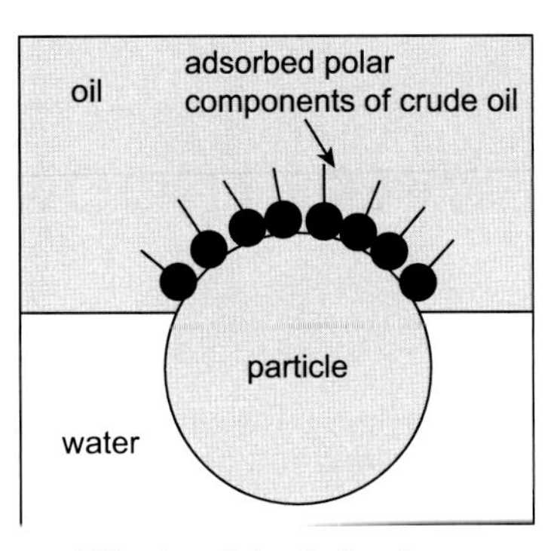

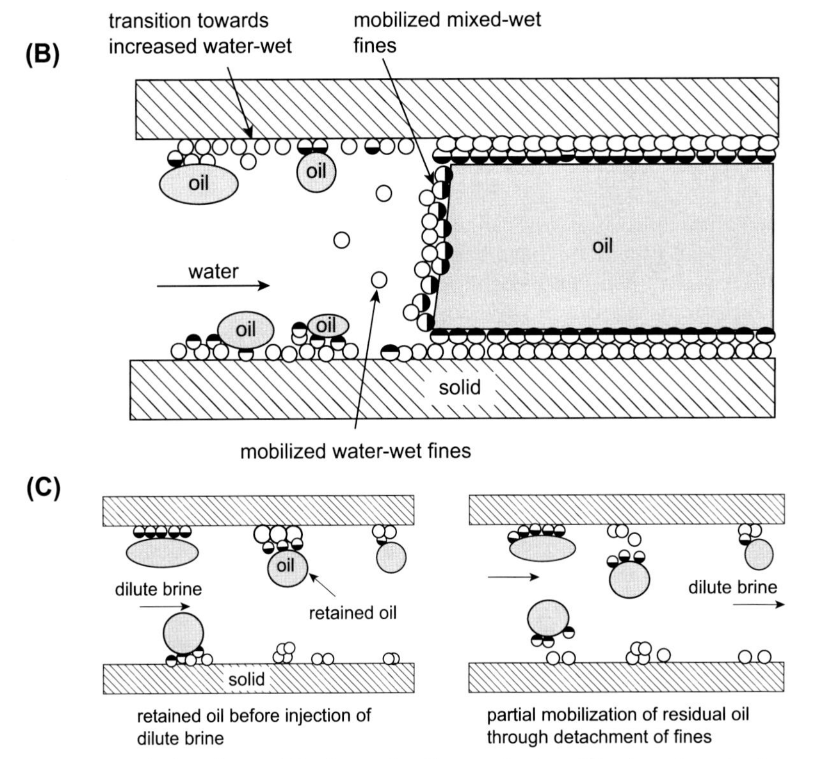

FIG. 2.1 The schematic description of the fines migration mechanism: **(A)** adsorption of polar components of crude oil to form mixed-wet fines; **(B)** partial stripping of mixed-wet fines from pore wall during waterflooding; and **(C)** mobilization of trapped oil. (Credit: From Tang, G.-Q., & Morrow, N.R. (1999). Influence of brine composition and fines migration on crude oil/brine/rock interactions and oil recovery. *Journal of Petroleum Science and Engineering, 24*(2), 99–111. https://doi.org/10.1016/S0920-4105(99)00034-0.)

increase as a mechanism of LSWF through coreflooding experiments. The study asserted that the observations of LSWF are similar to those of alkaline and surfactant floods, at least, in core-scaled experiments. It is explained that LSWF produces changes in reservoir fluid, fluid/rock interactions, and wettability. In the coreflood system, LSWF potentially results in the generation of OH^- from mineral reaction and the pH of the system increases. Surfactants are in situ generated from the residual oil at elevated pH. The study formulated a hypothetical mechanism of in situ generation of surfactant by pH increase and validated the mechanism with the coreflooding observations. The coreflooding observations of Tang and Morrow (1999) are referred to supplement the proposed mechanism. In the observations, the pH increases from 8 to 10 when injecting brine is changed from the formation brine with 15,150 ppm TDS to the low-salinity water with 1515 ppm TDS. It is concluded that the acid or polar components in the crude oil are saponified by the generalized reactions of Eqs. (2.1) and (2.2), and then, surfactants are in situ generated from the reactions. The in situ generated surfactant alters wettability and reduces IFT between water and oil. In addition, they act as emulsifying agents to disperse oil into the water. In the study, it is documented that an additional advantage of low salinity prevents a precipitation of the surfactant because of low concentration of divalent cations. The study summarized the following prerequisites for the successful LSWF: (1) acidic components in crude oil; (2) water-sensitive minerals; (3) initial water saturation; and (4) injecting low-salinity water with less than 5000 ppm.

$$(RCOO)_3C_3H_5 + 3NaOH \rightarrow 3(RCOONa) + C_3H_5(OH)_3 \tag{2.1}$$

$$2(RCOONa) + Ca(HCO_3)_2 \rightarrow (RCOO)_2Ca + 2NaHCO_3 \tag{2.2}$$

Multicomponent Ionic Exchange

Lager, Webb, Black, Singleton, and Sorbie (2008) reviewed the previous mechanisms and devised a new mechanism to explain the experimental observations of LSWF in sandstone reservoirs. The study carried out an effluent analysis of coreflood injecting low-salinity and high-salinity waters and investigated water chemistry. It showed a drop of Ca^{2+} and Mg^{2+} concentrations being lower than their concentrations in injecting brines (Fig. 1.7). These results indicate the strong adsorption of Ca^{2+} and Mg^{2+} onto the rock matrix. Based on these observations, a multicomponent ionic exchange (MIE) mechanism is proposed in sandstone. According to Arnarson and Keil (2000) and Sposito (1989), organic matters are possible to adhere onto clay minerals depending on the organic function of the organic matter and the clay surface conditions. The eight potential mechanisms of the adsorption exist as described in Table 2.1. The four mechanisms including cation exchange, ligand bonding, cation bridging, and water briding are strongly affected by the cation exchange (Fig. 2.2). It is explained that the cation exchange occurs during LSWF and it influences the four adsorption mechanisms of organic materials. The multivalent cations at a clay surface bond to polar compounds of crude oil (resin and asphaltene) forming organometallic complexes. The adsorbed complexes on the clay surface lead to oil-wetness of rock surface. Simultaneously, some organic polar compounds substitute the most labile cations and directly bond to the mineral surface. This substitution also promotes the initial oil-wetness of the clay surface. When LSWF is applied to the oil-wet sandstone rocks, the MIE occurs and replaces both organic polar compounds and organometallic complexes by uncomplexed cations on the clay surface. This study concluded that the replacement by MIE modifies the wettability of sandstone reservoirs toward water-wet and increases oil recovery.

TABLE 2.1
Mechanisms of Association Between Organic Functional Groups and Soil Minerals (Sposito, 1989)

Mechanisms	Organic Functional Group Involved
Cation exchange	Amino, ring NH, heterocyclic N (aromatic ring)
Protonation	Amino, heterocyclic N, carbonyl carboxylate
Anion exchange	Carboxylate
Water bridging	Amino, carboxylate, carbonyl, alcoholic OH
Cation bridging	Carboxylate, amines, carbonyl, alcoholic OH
Ligand bridging	Carboxylate
Hydrogen bonding	Amino, carbonyl, carboxyl, phenolic OH
Van der Waals interaction	Uncharged organic units

Salting-In Effect

Salting-out or salting-in effects have been used to describe the solubility of polar organic material in water as a function of salinity or ionic composition in the area of chemistry. RezaeiDoust, Puntervold, Strand, and Austad (2009) applied this theory to explain the observations of LSWF experiments and proposed the salting-in effect as another mechanism of LSWF. In the theory, organic compounds in water are solvated by the water structure, which is made by hydrogen

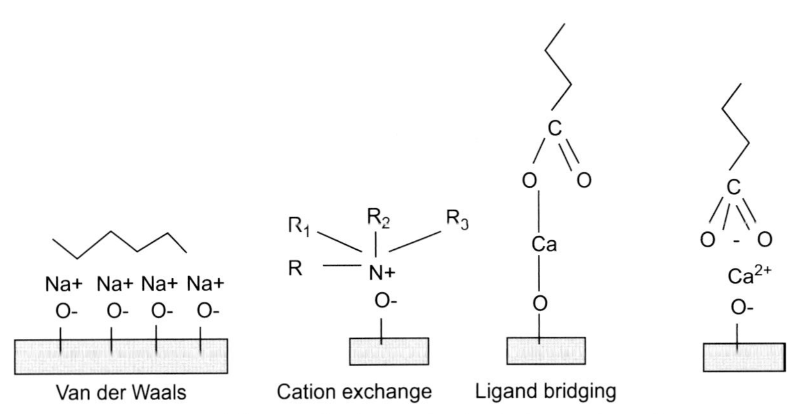

FIG. 2.2 Schematic description of diverse adhesion mechanism between clay surface and crude oil. (Credit: From Lager, A., Webb, K. J., Black, C. J. J., Singleton, M., & Sorbie, K. S. (2008). Low salinity oil recovery - an experimental investigation. *Petrophysics, 49*(1), 28–35. https://doi.org/SPWLA-2008-v49n1a2.)

bonds around the hydrophobic part. Inorganic cations of Ca^{2+}, Mg^{2+}, and Na^{+} destroy the water structure around the organic material and decrease the solubility of organic material in water. The divalent cations have the higher energy to break the water structure over monovalent cations. Therefore, the divalent cations have the higher effect on the solubility of organic material in water. Following the theory, the solubility of organic component drastically increases as salinity decreases, that is, salting-in effect. Following this description, the salting-in mechanism is formulated in the application of LSWF in sandstone. During LSWF, the adsorbed organic component onto clay surface might be detached from clay and dissolves into water. Initially, the adsorbed organic material should bond weakly to the surface. Then, the increasing solubility of organic material in water leads to the desorption of organic material from the surface improving the water-wetness of reservoirs. In addition, it is also explained that the release of cations from the clay surface increases the pH as Eq. (2.3).

$$\text{Clay-Ca}^{2+} + H_2O \rightleftarrows \text{Clay-H}^{+} + Ca^{2+} + OH^{-} \tag{2.3}$$

Electrical Double Layer Expansion

Ligthelm et al. (2009) proposed the mechanism of electrical double layer (EDL) expansion to explain the observations of LSWF experiments. In sandstone reservoirs, clay minerals often have the negative electrical charge owing to crystal lattice. The crude oil also exhibits the negative charge in nature. The multivalent cations of Mg^{2+} and Ca^{2+} in formation water are believed to link between clay and crude oil as explained by Lager et al. (2008). In high salinity condition, the sufficient concentration of the cations screens the negative charge of both clay and crude oil and suppresses the electrostatic repulsive force between clay and crude oil. This leads to the low level of the negative ζ-potential. Therefore, oil enables to react with these clay particles forming organometallic complexes, leading to the local oil-wetness of clay surface. In the low salinity condition, i.e., lower level of ionic strength, the screening potential decreases with a reduction of concentration of the multivalent cations and increases the ζ-potential. It expands the electrical double layers increasing the electrostatic repulsion between clay and crude oil. It is believed that the repulsive forces exceed the binding forces via the multivalent cation bridge, and then, the oil particles can be detached from the clay surfaces. It leads to the wettability modification toward water-wetness. This study also explained the observation of fines migration (Tang & Morrow, 1999) with the EDL expansion. When the ionic strength is reduced further, the mutual repulsive electrostatic forces within the clay minerals potentially exceed binding forces and yield stripping of oil-bearing fines from the pore wall.

pH Increase

Whereas McGuire et al. (2005) explained the pH increase to trigger the in situ generation of surfactant, Austad, Rezaeidoust, and Puntervold (2010) suggested the pH increase as the mechanism of LSWF. Austad et al. (2010) explained the physical process in terms of pH and basic and acidic organic components, not in situ saponification. In an actual reservoir condition, the basic and acidic organic components of oil can adsorb onto the clay representing oil-wet system. The clay mineral is a cation exchanger and has a relatively large surface area. Initially, the basic and acidic organic materials bond to the clay with inorganic cation, mainly Ca^{2+}, from the formation brine. When the low-salinity water is injected into the reservoir, it dilutes the ionic concentration close to clay surface and an equilibrium associated with the geochemical interaction of brine-rock is disturbed. As a result, it causes the desorption of Ca^{2+} and the coadsorption of H^+ replacing the desorption of Ca^{2+} as shown in Eq. (2.3). This exchange causes a local increase in pH close to clay. Successively, the increasing OH^- close to the clay surface leads to the reactions with the adsorbed basic and acidic materials as in an ordinary acid-base proton transfer reactions of Eqs. (2.4) and (2.5). These reactions result in the desorption of basic and acidic organic materials from the clay modifying wettability toward water-wetness (Fig. 2.3).

$$\text{Clay-NHR}_3^+ + OH^- \rightleftarrows \text{Clay} + R_3N + H_2O \tag{2.4}$$

$$\text{Clay-RCOOH} + OH^- \rightleftarrows \text{Clay} + RCOO^- + H_2O \tag{2.5}$$

MECHANISMS IN CARBONATE RESERVOIRS

Previous experiments of coreflooding and a pilot test have indicated that improved oil recovery by LSWF is mainly attributed to the wettability modification in mixed- to oil-wet sandstone. A number of mechanisms are suggested to describe the enhanced oil recovery of LSWF in sandstone reservoirs, and the clay has the key role to explain the mechanism. Because carbonates have negligible clay mineral content, the proposed mechanisms of LSWF in sandstone reservoirs are not appropriate to the carbonate reservoirs. In addition, the structure and mineralogy of the carbonate reservoirs are different to the sandstone reservoirs. Therefore, other mechanisms of LSWF or smart waterflood in carbonate reservoirs have been proposed, and they are briefly described in this section.

Potential-Determining Ions

Zhang, Tweheyo, and Austad (2007) proposed the mechanism that the adsorptions of potential-determining ions (Ca^{2+}, Mg^{2+}, and SO_4^{2-}) onto the chalk surface contribute to the wettability modification of chalk surface toward water-wetness. Initially, the negatively charged organic component of crude oil adheres to the positively charged surface of chalk. Generally, the formation water has negligible SO_4^{2-}, but the seawater has considerable concentration of SO_4^{2-}. When the seawater is injected into the oil-wet chalk reservoir, the SO_4^{2-} in injecting seawater potentially adsorbs onto the positively charged water-wet sites on the chalk. The adhesion of SO_4^{2-} on the chalk surface lowers the degree of positive charge density of chalk surface. The reduction in the positive surface charge density of chalk surface will promote to decrease the electrostatic repulsion between cation and chalk surface. The less electrostatic repulsion, the more Ca^{2+} enables to adsorb onto the chalk and the more Ca^{2+} remains close to the chalk surface. It also accelerates the coadsorption of SO_4^{2-} on the surface. The Ca^{2+} potentially reacts with the adsorbed carboxylic groups, which are the negatively charged organic components of crude oil. The reaction releases the organic carboxylic

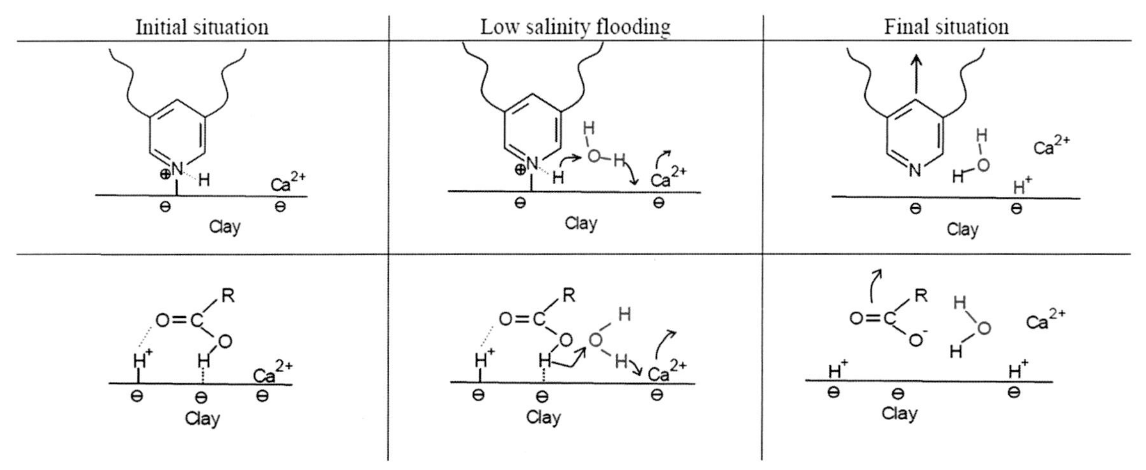

FIG. 2.3 The schematic description of pH increase mechanism. (Credit: From Austad, T., Rezaeidoust, A., & Puntervold, T. (2010). Chemical mechanism of low salinity water flooding in sandstone reservoirs. In *Paper presented at the SPE improved oil recovery Symposium, Tulsa, Oklahoma, USA, 2428 April*. https://doi.org/10.2118/129767-MS.)

materials from chalk surface (Fig. 2.4). In high temperature conditions, another mechanism is formulated with the substitution of Ca^{2+} by Mg^{2+} onto chalk surface. The substitution potentially displaces carboxylic groups reacted with Ca^{2+} on the chalk surface and increases the water-wetness of chalk surface. In addition, the coadsorption of SO_4^{2-} lowers the electrostatic repulsive force. In the excess of SO_4^{2-} on the chalk surface, the ionic interaction between Mg^{2+} and SO_4^{2-} increases the concentration of Mg^{2+} close to the chalk surface (Fig. 2.4). The increasing concentration of Mg^{2+} close to the surface potentially leads to the more substitution of Ca^{2+} by Mg^{2+}. Following the mechanism, neither Ca^{2+} nor Mg^{2+} can remove negatively charged carboxylic organic material from the positively charged chalk surface without SO_4^{2-} in both low and high temperatures.

Mineral Dissolution

Hiorth, Cathles, and Madland (2010) explored whether rock surface charge or mineral dissolution is attributed to the wettability modification of LSWF. Based on experimental results, the study numerically constructed chemical model to determine the potential mechanism. The chemical model predicts the ζ-potential of calcite and the adsorption of SO_4^{2-} on surface. The predictions are compared with the experimental results of the studies (Strand, Høgnesen, & Austad, 2006; Thompson & Pownall, 1989; Zhang & Austad, 2006). The predictions by chemical model are comparable with the experimental observations of ζ-potential of calcite and the adsorption of SO_4^{2-}. Using the chemical model, Hiorth et al. (2010) concluded that the change of ζ-potential by water chemistry hardly explains the observations of temperature-dependent oil recovery, which are observed in the studies (Zhang & Austad, 2006; Zhang, Tweheyo, & Austad, 2006, 2007). Therefore, the study proposed that the mechanism of calcite mineral dissolution is accountable for the increasing oil recovery and temperature dependency of oil recovery during LSWF. The calcite mineral dissolution can explain the temperature-dependent observations of experiments. At a lower temperature condition, seawater is equilibrium with calcite. When the temperature increases, Ca^{2+} is started to react with SO_4^{2-} and anhydrite is precipitated. The loss of Ca^{2+} is compensated from the calcite dissolution. When the calcite dissolves, the adsorbed oil onto the calcite surface is released and wettability is changed (Fig. 2.5).

Surface Charge

Alotaibi, Nasr-El-Din, and Fletcher (2011) investigated the electrokinetics of limestone and dolomite suspensions at different temperature conditions (25°C and 50°C). Using the synthetic formation brine, seawater, and aquifer water, the study measured the ζ-potentials of limestone and dolomite suspensions using phase analysis light scattering (PALS) techniques. The interfacial phenomena at the surface between rock and brine is

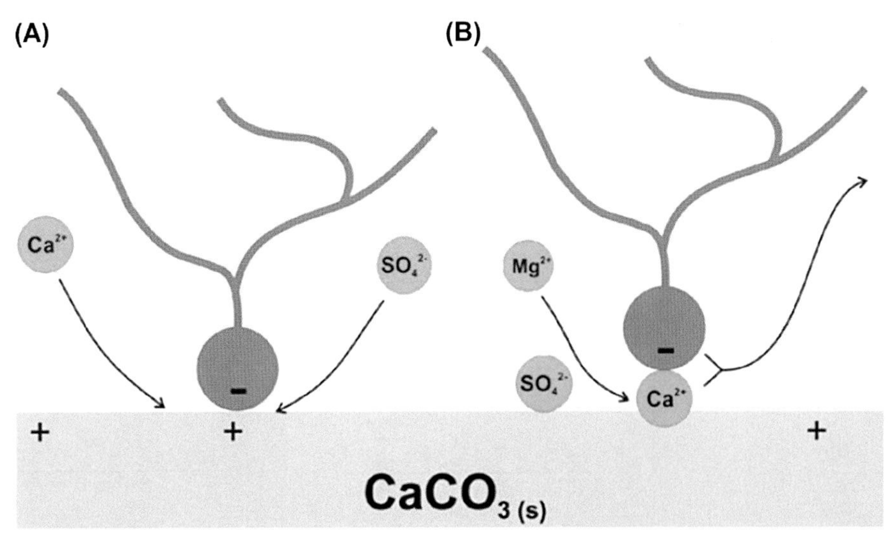

FIG. 2.4 The schematic descriptions of wettability modification by potential-determining ions at low and high temperatures. (Credit: From Zhang, P., Tweheyo, M.T. & Austad, T. (2007). Wettability alteration and improved oil recovery by Spontaneous Imbibition of seawater into chalk: Impact of the potential determining ions Ca^{2+}, Mg^{2+}, and SO_4^{2-}. *Colloids and Surfaces A: Physicochemical and Engineering Aspects, 301*(1), 199–208. https://doi.org/10.1016/j.colsurfa.2006.12.058.)

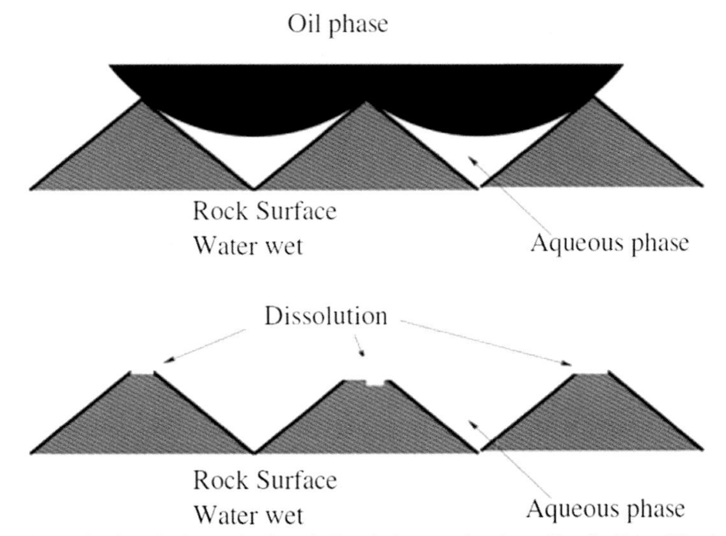

FIG. 2.5 The schematic descriptions of mineral dissolution mechanism. (Credit: From Hiorth, A., Cathles, L. M., & Madland, M. V. (2010). The impact of pore water chemistry on carbonate surface charge and oil wettability. *Transport in Porous Media, 85*(1), 121. https://doi.org/10.1007/s11242-010-9543-6.)

controlled by the EDL forces. The ζ-potential is the key indicator to state EDL. Generally, ζ-potential is a strong function of pH and ionic strength, and the crude oil tends to have negative surface charge because of the carboxylate groups of the crude oil. In the experimental study, it is observed that low-salinity water results in more negative surface charge on the limestone and dolomite particles. Because both oil and limestone/dolomite have the negative surface charges, oil hardly adheres to the rock surface resulting in water-wetness. Therefore, this study hypothesized that the change of surface potential in carbonate rock from positive to negative alters the wettability of carbonate rock toward water-wetness.

Adhesion Energy of Crude Oil/Brine/Rock

A number of studies have exploited the adhesion energy of the crude oil/brine/rock system to explain the experimental observations in carbonate reservoirs. The studies (Brady, Krumhansl, & Mariner, 2012; Brady and Thyne 2016) have proposed the concept of fractional wettability, which describes the process of indirect and direct adhesion of oil to carbonate reservoir. The balance of forces (van der Waals force, structural force of hydration or hydrogen-bonding forces, and electrostatic force) determines the status of oil in the system of mineral surface and fluids. The oil can be free phase or directly and indirectly adhered to the mineral surface in the reservoir. The indirect adherence of oil on the mineral surface is accomplished through a three-layer oil/water/rock configuration (Fig. 2.6A). It is a reversible process. The degree of oil sorption is a function of the distance of separation between the oil and mineral, i.e., thickness of the water layer. The thickness of water layer depends on the net charge present at the two interfaces of rock/water and oil/water. The charge is sensitive to the ionic composition of water layer. In addition, a fraction of oil can directly adhere to the minerals without water layer (Fig. 2.6B). When the attractive force is greater than the disjoining pressure, the direct adsorption of oil occurs (Buckley & Liu, 1998; Hirasaki, 1991). The disjoining pressure indicates the force that tends to disjoin or separate the two interfaces. This study assumed that the directly adhered oil can be removed only by the external treatment such as surfactant injection, alkaline injection, etc. In addition, there is an incomplete or extremely slow desorption of species, i.e., ions and macromolecules, from the mineral surfaces. This is because the directly sorbed species forms the much stronger anhydrous bonds by linking directly to the high energy sites on the mineral surfaces. The concept of the functional wettability distinguishes the indirect and direct adhesions separately to identify the causes of wettability alteration in carbonate rocks. The indirect adhesion of oil on the rock surface is the main target for the wettability modification by LSWF. Using the concept, a predictive model is developed to bridge the experimental observations and the modifications of oil and mineral surface charges by changes in water chemistry. The model requires the modeling of both calcite and oil surface charges.

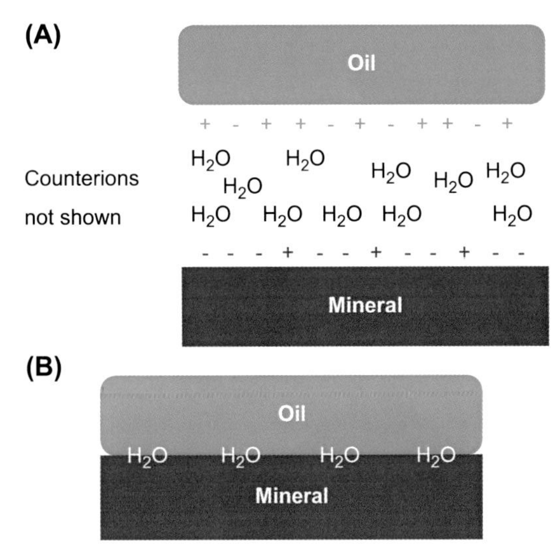

FIG. 2.6 **(A)** Indirectly adhered oil through a three-layer oil/water/rock and **(B)** directly adhered oil on the rock. (Credit: Brady, P.V., & Thyne, G. (2016). Functional wettability in carbonate reservoirs. *Energy and Fuels, 30*(11), 9217–9225. https://doi.org/10.1021/acs.energyfuels.6b01895.)

The carbonate surfaces have both cation- and anion-exchange sites. Crude oil has cationic surface groups of $-NH^+$ and $-COOCa^+$, which links to the cation-exchange site of carbonate surface, and anionic surface group of $-COO^-$, which links to anion-exchange site of carbonate surface. In addition, the carbonate surface consists of the hydrated calcium and carbonate sites. The surface charge involved with the calcite surface is developed through the sites gaining or losing hydrogen ions, i.e., surface acid-base reactions, and adsorption of multivalent cations or anions on the charged sites. Then, the layer of the hydrated counterions is formed around the charged mineral surface to balance the surface charge. Therefore, the EDL forms at the interface of mineral and water. At the interface of oil and water, another EDL generates. Conventionally, the oil and mineral surface charges are referred by measuring ζ-potential. However, this study suggested the development of the predictive model, which employs surface complexation models of the oil and carbonate interfaces to describe the surface mechanisms quantitatively. The complexation models are constructed using a diffuse layer model of the EDL. Then, the indirect adhesion of oil on the carbonate surface is formulated through bridging between the two complexation models of the oil and carbonate interfaces.

Another study by Chen et al. (2018) formulated the interrelated mechanisms, consequently determining the adhesion energy of COBR based on the experimental results and theoretical modeling. The main hypothesis is that increasing oil recovery is attributed to the lower adhesion energy between crude oil and carbonate rock across the brine. As a result, the lower adhesion energy releases the crude oil from the surface and enables the crude oil to flow throughout the reservoir. This study suggested that three different mechanisms are interrelated and, synergistically, influence the adhesion energy of COBR by reducing ionic strength. The three different mechanisms are (1) changes to the colloidal interaction forces; (2) roughening, dissolution, and restructuring of the underlying calcite surfaces; and (3) removal of preadsorbed organic-ionic layers (ad-layers) as flakes that carry with oil.

The first mechanism is related to the colloidal interaction forces: (1) EDL repulsion; (2) van der Waals; and (3) hydration (structural) forces. The adhesion energy is the result of the interaction forces between crude oil/brine and brine/rock interfaces across the brine film. These forces are collectively given by the extended Derjaguin-Landau-Verwey-Overbeek (DLVO) theory. The extended DLVO theory calculates the adhesion energy, and Eqs. (2.6) and (2.7) express the extended DLVO between two spherical particles. The relationship between the adhesion energy and wettability is quantified with the Young-Dupré equation of Eq. (2.8), which calculates the contact angle corresponding to the adhesion energy. Following the extended DLVO theory, adhesion energy decreases until ionic strength decreases and reaches to the critical ionic strength. With the Young-Dupré equation, the decrease in the adhesion energy results in decreasing contact angle and increasing water-wetness (Fig. 2.7). This result is attributed to the relatively higher EDL repulsive force over other forces. However, the further reduction in ionic strength below the critical ionic strength results in the increasing adhesion energy. This reversal trend across the critical ionic strength is responsible for the effect of brine concentration at a specific distance on the EDL repulsion force. The concentration of brine less than critical ionic strength makes negligible EDL contribution. It implies optimum ionic concentration to maximize the wettability modification toward water-wetness and, consequently, oil recovery.

$$W(d) = \frac{64k_BT}{\kappa}C\left[\tanh\left(\frac{ze\psi_0}{4k_BT}\right)\right]^2\exp(-\kappa d) - \frac{A}{12\pi(d+\Delta d)^2} + \left(\frac{\delta}{d}\right)^{20} \tag{2.6}$$

where W is the colloidal interaction force; κ^{-1} is the Debye length; k_B is the Boltzmann's constant; T is the temperature; C is the brine concentration; z is the brine electrolyte valence; e is the elementary electronic charge;

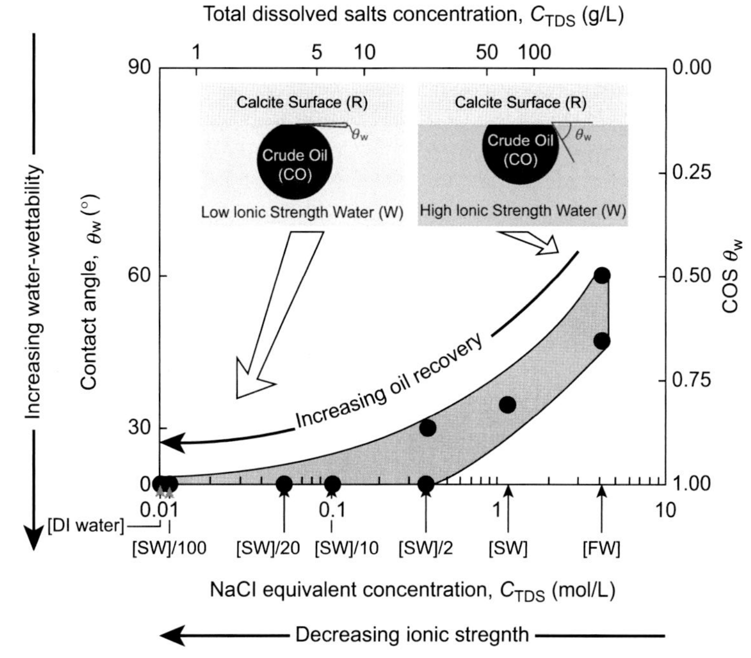

FIG. 2.7 The effect of salinity on the contact angle and wettability. (Credit: From Chen, S.-Y., Kaufman, Y., Kristiansen, K., Dobbs Howard, A., Cadirov Nicholas, A., Seo, D., & et al. (2018). New Atomic to molecular scale insights into smart water flooding mechanisms in carbonates. In *Paper presented at the SPE improved oil recovery Conference, Tulsa, Oklahoma, USA, 1418 April*. https://doi.org/10.2118/190281-MS.)

ψ_0 is the surface potential; d is the distance; A and $\Delta d$are the Hamaker constant and offset distance, respectively, for the van der Waals attraction; and δ is the "hard-wall" hydration repulsion layer thickness.

$$\kappa^{-1} = \sqrt{\frac{\varepsilon_0 \varepsilon_r k_B T}{2Ce^2}} \tag{2.7}$$

where ε_r is the dielectric constant and ε_0 is the permittivity in free space.

$$W^o_{o/w/r} = \gamma_{o/w}\left(1 - \cos\theta^o_w\right) \tag{2.8}$$

where $W^o_{o/w/r}$ and θ^o_w are the intrinsic adhesion energy and intrinsic contact angle, respectively, on molecularly smooth, nonstructured, and flat surfaces and $\gamma_{o/w}$ is the interfacial tension between oil and water.

The proposed mechanisms of LSWF with the potential-determining ions, EDL expansion, and surface charge change can be explained with the first mechanism using extended DLVO theory and Young-Dupré equation. However, they ignore the changes in (nanoscale to microscale) surface roughness and roughening, which can significantly influence on the overall adhesion energy. The extended DLVO theory and Young-Dupré equation, conventionally, are strictly appropriate to molecularly smooth and flat surface. Typical reservoirs show complex surface, not flat, and become further roughened because of mineral dissolution during LSWF. The second mechanism on the roughening of the underlying calcite surface overcomes the limitations, which are involved with the first mechanism. The initial surface roughness and more roughening due to mineral dissolution on the adhesion energy are included in the modified Young-Dupré equation of Eq. (2.9), which introduces the concept of an effective adhesion energy. Notably, the mineral dissolution roughens the calcite surface and decreases contact angle improving wettability toward water-wetness.

$$W_{\text{effective},\,o/w/r} = \phi_{o/r} W^o_{o/w/r} = \gamma_{o/w}\left(1 - \cos\theta^o_w\right) \tag{2.9}$$

where $W_{\text{effective},\,o/w/r}$ is the effective adhesion energy and $\phi_{o/r}$ is a surface roughness factor, defined as the ratio between real contact area of oil and rock across the aqueous brine film and the projected, flat contact area on the smooth surface.

In addition to the physical-chemical considerations, the third mechanism takes account of the removal of preadsorbed organic-ionic ad-layer, i.e., flakes, from the calcite surface. During LSWF, the flakes are released from calcite surface and suspended in the brine. The detachment of organic materials with the flakes potentially increases oil production. In addition, the removal of the flakes leaves behind less organic materials on calcite surfaces and it makes the underlying hydrophilic calcite to be exposed. Consequently, the removal of preadsorbed organic-ionic ad-layer because of low ionic strength promotes the wetness toward water-wet. The study concluded that the increasing oil recovery via the LSWF is attributed to the three different but interrelated mechanisms.

COMPARISON OF LSWF BETWEEN SANDSTONE AND CARBONATE RESERVOIRS

Extensive mechanisms have been proposed to explain the enhanced oil recovery of LSWF for both sandstone and carbonate reservoirs. Majority of the proposed mechanisms have explained that physical and/or chemical phenomena are attributed to wettability alteration, consequently improving oil recovery. Because of the striking differences (clay content and surface potential) between sandstone and carbonate reservoirs, different physical and/or chemical reactions have been proposed to explain the wettability modification underlain in sandstone and carbonate reservoirs, respectively. Carbonate reservoirs could contain slight clay, but it is not much and hardly exposed to the invading fluid. Sandstone could have a wide range of clay content. In addition, inherent positively-charged carbonate and negatively-charged quartz and clay lead to different potential reactions contributing to wettability alteration. The positively charged calcite surface can attract the carboxylic organic component directly, and negatively charged clay surface mostly needs cations to bind the organic materials. Different mineralogy and surface potential require different conditions to explain the wettability modification effect by LSWF for carbonate and sandstone reservoirs. Following section summarizes these conditions.

Conditions in Sandstone

- Sandstone has negatively charged surface because of the abundant clay. The clay minerals are the hydrous aluminous phyllosilicates and include the several types of minerals. The clay minerals have the negative charges occurring cation exchange, and kaolinite and illite usually have higher capacity of cation exchange.
- Crude oil has the basic and/or acidic polar components. The acidic material usually has the functional group, the carboxylic group (-COOH). The basic component with high polarity has nitrogen in the aromatic molecules (R_3N). Conventionally, the acidic and basic materials at the oil-water interface undergo the fast proton exchange reactions, which are sensitive to the pH of brine. The reactions at the interface are described in Eqs. (2.10) and (2.11). The species of R-COOH and R_3NH^+ have the highest affinity to the negatively charged clay minerals.

$$\text{R-COOH} + H_2O \rightleftarrows \text{R-COO}^- + H_3O^+ \quad (2.10)$$

$$R_3N: + H_2O \rightleftarrows R_3NH^+ + OH^- \quad (2.11)$$

- Formation water must be present and has divalent cations (Ca^{2+} and Mg^{2+}).
- Low-salinity water to be injected has salinity in the range of 1000–2000 ppm TDS. LSWF effects have been observed with the salinity up to 5000 ppm TDS.
- There appears to be no effective temperature conditions.
- The produced water usually shows the increase in pH. However, there is a controversy about whether the pH increase is a necessary condition or just experimental observation.
- Experiments reported the effects of LSWF with and without production of fines.
- Increasing pressure of injector is observed when injecting brine is switched from high-salinity water to low-salinity water.

Conditions in Carbonate Rocks

- The calcite surface has positively charged potential because of complex molecular structure.
- Crude oil must have the acidic polar component. The acidic polar component generates the negatively charged carboxylic materials, R-COO^-, as shown in Eq. (2.10). The negatively charged component adsorbs on the positively charged carbonate surface. The adsorption, initially, results in mixed- to oil-wetness of carbonate reservoir. In terms of natural bases of crude oil, they hardly coadsorb with the negatively charged carboxylic material on the chalk surface because of steric hindrance. Rather than, the acid-base complexes form and they reduce the adsorption of the carboxylic material implying less oil-wetness.
- Low-salinity water to be injected must have SO_4^{2-} and Ca^{2+} and/or Mg^{2+}. The limitation regarding overall salinity is not reported.
- The effective temperature conditions are reported with the range of 70–130°C.

REFERENCES

Alotaibi, M. B., Nasr-El-Din, H. A., & Fletcher, J. J. (2011). Electrokinetics of limestone and dolomite rock particles. *SPE Reservoir Evaluation and Engineering, 14*(5), 594–603. https://doi.org/10.2118/148701-PA.

Arnarson, T. S., & Keil, R. G. (2000). Mechanisms of pore water organic matter adsorption to montmorillonite. *Marine Chemistry, 71*(3), 309–320. https://doi.org/10.1016/S0304-4203(00)00059-1.

Austad, T., Rezaeidoust, A., & Puntervold, T. (2010). Chemical mechanism of low salinity water flooding in sandstone reservoirs. In *Paper presented at the SPE improved oil recovery symposium, Tulsa, Oklahoma, USA, 24–28 April*. https://doi.org/10.2118/129767-MS.

Brady, P. V., Krumhansl, J. L., & Mariner, P. E. (2012). Surface complexation modeling for improved oil recovery. In *Paper presented at the SPE improved oil recovery symposium, Tulsa, Oklahoma, USA, 14–18 April*. https://doi.org/10.2118/153744-MS.

Brady, P. V., & Thyne, G. (2016). Functional wettability in carbonate reservoirs. *Energy and Fuels, 30*(11), 9217–9225. https://doi.org/10.1021/acs.energyfuels.6b01895.

Buckley, J. S., & Liu, Y. (1998). Some mechanisms of crude oil/brine/solid interactions. *Journal of Petroleum Science and Engineering, 20*(3), 155–160. https://doi.org/10.1016/S0920-4105(98)00015-1.

Chen, S.-Y., Kaufman, Y., Kristiansen, K., Dobbs Howard, A., Cadirov Nicholas, A., Seo, D., et al. (2018). New atomic to molecular scale insights into smart water flooding mechanisms in carbonates. In *Paper presented at the SPE improved oil recovery conference, Tulsa, Oklahoma, USA, 14–18 April*. https://doi.org/10.2118/190281-MS.

Hiorth, A., Cathles, L. M., & Madland, M. V. (2010). The impact of pore water chemistry on carbonate surface charge and oil wettability. *Transport in Porous Media, 85*(1), 121. https://doi.org/10.1007/s11242-010-9543-6.

Hirasaki, G. J. (1991). Wettability: Fundamentals and surface forces. *SPE Formation Evaluation, 6*(02), 217–226. https://doi.org/10.2118/17367-PA.

Lager, A., Webb, K. J., Black, C. J. J., Singleton, M., & Sorbie, K. S. (2008). Low salinity oil recovery - an experimental investigation. *Petrophysics, 49*(1), 28–35. SPWLA-2008-v49n1a2.

Ligthelm, D. J., Gronsveld, J., Hofman, J., Brussee, N., Marcelis, F., & van der Linde, H. (2009). Novel waterflooding strategy by manipulation of injection brine composition. In *Paper presented at the EUROPEC/EAGE conference and exhibition, Amsterdam, The Netherlands, 8–11 June*. https://doi.org/10.2118/119835-MS.

McGuire, P. L., Chatham, J. R., Paskvan, F. K., Sommer, D. M., & Carini, F. H. (2005). Low salinity oil recovery: An exciting new EOR opportunity for Alaska's North Slope. In *Paper presented at the SPE Western Regional Meeting, Irvine, California, 30 March 1 April*. https://doi.org/10.2118/93903-MS.

RezaeiDoust, A., Puntervold, T., Strand, S., & Austad, T. (2009). Smart water as wettability modifier in carbonate and sandstone: A discussion of similarities/differences in the chemical mechanisms. *Energy and Fuels, 23*(9), 4479–4485. https://doi.org/10.1021/ef900185q.

Sposito, G. (1989). *The chemistry of soils*. New York: Oxford University Press.

Strand, S., Høgnesen, E. J., & Austad, T. (2006). Wettability alteration of carbonates—effects of potential determining ions (Ca^{2+} and SO_4^{2-}) and temperature. *Colloids and Surfaces A: Physicochemical and Engineering Aspects, 275*(1), 110. https://doi.org/10.1016/j.colsurfa.2005.10.061.

Tang, G.-Q., & Morrow, N. R. (1999). Influence of brine composition and fines migration on crude oil/brine/rock interactions and oil recovery. *Journal of Petroleum Science and Engineering, 24*(2), 99–111. https://doi.org/10.1016/S0920-4105(99)00034-0.

Thompson, D. W., & Pownall, P. G. (1989). Surface electrical properties of calcite. *Journal of Colloid and Interface Science, 131*(1), 74–82. https://doi.org/10.1016/0021-9797(89)90147-1.

Zhang, P., & Austad, T. (2006). Wettability and oil recovery from carbonates: Effects of temperature and potential determining ions. *Colloids and Surfaces A: Physicochemical and Engineering Aspects, 279*(1), 179–187. https://doi.org/10.1016/j.colsurfa.2006.01.009.

Zhang, P., Tweheyo, M. T., & Austad, T. (2006). Wettability alteration and improved oil recovery in Chalk: the effect of calcium in the presence of sulfate. *Energy and Fuels, 20*(5), 2056–2062. https://doi.org/10.1021/ef0600816.

Zhang, P., Tweheyo, M. T., & Austad, T. (2007). Wettability alteration and improved oil recovery by spontaneous imbibition of seawater into chalk: Impact of the potential determining ions Ca^{2+}, Mg^{2+}, and SO_4^{2-}. *Colloids and Surfaces A: Physicochemical and Engineering Aspects, 301*(1), 199–208. https://doi.org/10.1016/j.colsurfa.2006.12.058.

CHAPTER 3

Modeling of Low-Salinity and Smart Waterflood

ABSTRACT

This chapter discusses the numerical simulations of low-salinity waterflood (LSWF) incorporating the proposed mechanisms for sandstone and carbonate reservoirs. Because the geochemical reactions including aqueous reaction, mineral reaction, ion exchange, and surface complexation and electrokinetics in the crude oil/brine/rock system are responsible for underlying mechanisms, they are implemented in the numerical simulations. In addition, a number of approaches have been proposed to model the mechanism mimicking the historical results of experimental works and a couple of field trial tests. This chapter discusses the important features of geochemistry to support the numerical modeling of LSWF process. It also describes a variety of numerical studies, which have developed the empirical and mechanistic modeling of LSWF process both in sandstone and carbonate reservoirs.

In the previous chapter, the various mechanisms of low-salinity waterflood (LSWF) and smart waterflood have been formulated based on the experimental works and field-scaled tests. Because of the different mineralogy and surface charge between sandstone and carbonate reservoirs, different mechanisms have been proposed in sandstone and carbonate reservoirs. Understanding the mechanisms of LSWF and smart waterflood is essential to the development of numerical modeling of LSWF and smart waterflood. Extensive attempts for the numerical modeling include the comprehensive geochemical reactions, electrokinetics, and empirical and mechanistic modeling of LSWF mechanism, mainly wettability modification. This section presents the geochemistry, which contributes to the proposed mechanisms of LSWF and smart waterflood. The detail explanations of geochemistry could be found in the references (Appelo & Postma, 1999; Bethke, 1996; Drever, 1997). Following sections describe the empirical and mechanistic modeling incorporating the mechanisms of LSWF and smart waterflood in sandstone and/or carbonate reservoirs.

GEOCHEMISTRY

Equilibrium Thermodynamics

In a natural system, the complete chemical equilibrium is rarely achieved, especially, where biological processes are involved (Drever, 1997). However, the calculation of equilibrium confidently approximates the real equilibrium thermodynamics and indicates the direction of the chemical reactions of systems. The Gibbs free energy of the system is a useful tool to indicate the status of the chemical reactions. For a system at constant pressure and temperature, the Gibbs free energy is defined with temperature, pressure, and enthalpy as shown in Eq. (3.1).

$$G = U + pV - TS = H - TS \tag{3.1}$$

where G is the Gibbs free energy, U is the internal energy, p is the pressure, V is volume, H is the enthalpy, T is the temperature, and S is the entropy.

All chemical reactions spontaneously tend to reach the equilibrium in a state of minimum Gibbs free energy. A chemical reaction, not at equilibrium, releases the energy and moves toward the equilibrium state. The change of Gibbs free energy indicates whether the chemical reactions are under spontaneous process or already in the equilibrium state. The change in the Gibbs free energy is defined as in Eq. (3.2). Following the definitions of Gibbs free energy change and equilibrium state, a spontaneous process involves the negative change of Gibbs free energy. When the change of Gibbs free energy is zero, the system is at the equilibrium.

$$\Delta G = \Delta H - T\Delta S \tag{3.2}$$

where ΔG is the change in the Gibbs free energy, ΔH is the change in the enthalpy, and ΔS is the change in the entropy.

Hybrid Enhanced Oil Recovery using Smart Waterflooding. https://doi.org/10.1016/B978-0-12-816776-2.00003-9

In a multiphase and multicomponent system, the Gibbs free energy varies by species of the system. The partial molar Gibbs energy of the species can be defined using the theory of ideal solutions as shown in Eq. (3.3) and it is known as a chemical potential.

$$\mu_i = \left(\frac{\partial G}{\partial n_i}\right)_{T,p} = \mu_i^o + RT \ln a_i \quad (3.3)$$

where i indicates the species, μ_i is the chemical potential of species i, n_i is the number of moles of species, μ_i^o is a constant, which is the chemical potential of species i at a standard condition, R is the ideal gas constant, and a_i is the activity of the species i. The standard condition refers the pure state of the species i at the temperature of the solution and one atmospheric pressure. Following the definition, the equilibrium of reactions in the multiphase and multicomponent system ideally can be determined from the tabulated values of standard chemical potential.

Equilibrium constant

The fundamental description of geochemical equilibrium follows the law of mass action. The law of mass action can be applicable to any type of reactions: mineral dissolution, the complex formation between species, the dissolution of gases in water, etc. For the generalized reaction corresponding to Eq. (3.4), an equilibrium constant determines the distribution of species in the equilibrium state and it is a function of activity of species and stoichiometric coefficient following Eq. (3.5).

$$aA + bB \leftrightarrow cC + dD \quad (3.4)$$

$$K_{eq} = \frac{[C]^c[D]^d}{[A]^a[B]^b} \quad (3.5)$$

where K_{eq} is the equilibrium constant and $[i]$ is the activity of species i.

The equilibrium constant also can be derived from the expressions of chemical potential and Gibbs free energy. The difference between the Gibbs free energy of the products and that of the reactants, as shown in Eq. (3.6), can be rewritten in terms of chemical potential of Eq. (3.3). It is rearranged with Eq. (3.7) and equivalent to Eq. (3.8).

$$\Delta G = G_{\text{products}} - G_{\text{reactants}} \quad (3.6)$$

$$\Delta G = c\mu_C^o + d\mu_D^o - a\mu_A^o - b\mu_B^o + RT \ln\left(\frac{a_C^c a_D^d}{a_A^a a_B^b}\right) \quad (3.7)$$

$$\Delta G = \Delta G^o + RT \ln K_{eq} \quad (3.8)$$

where ΔG^o is the change in the standard Gibbs free energy of the reaction and μ^o is the standard chemical potential.

At the equilibrium, the change in Gibbs free energy is zero and the equilibrium constant can be calculated directly with the change in the standard Gibbs free energy following Eq. (3.9).

$$\ln K_{eq} = -\frac{\Delta G^o}{RT} \quad (3.9)$$

The equilibrium constant is highly sensitive to the temperature and less sensitive to the pressure (Appelo & Postma, 1999). The variations of equilibrium constant with temperature are usually calculated with van't Hoff equation as shown in Eq. (3.10) (Appelo & Postma, 1999).

$$\log K_{eq,T_1} - \log K_{eq,T_2} = \frac{-\Delta H}{2.303R}\left(\frac{1}{T_1} - \frac{1}{T_2}\right) \quad (3.10)$$

where T_1 and T_2 are the temperatures, and K_{eq,T_1} and K_{eq,T_2} are the equilibrium constants at temperatures T_1 and T_2.

Acitivity

In the calculation of the chemical potential and equilibrium constant, the acitivity of species is necessary to be defined. For an aqueous solution, the acitivity of ion can be thought as the effective concentration of ion and defined as in Eq. (3.11).

$$a_i = \gamma_i m_i \quad (3.11)$$

where γ_i is the activity coefficient of species i and m_i is the molality of species i.

In an infinite diluted solution, the activity coefficient of species approaches unity and the acitivity of a solute becomes its concentration following the definition of acitivity. Otherwise, Debye-Hückel theory calculates the activity coefficient. In electrolyte solutions, the theory allows an activity coefficient for a single ion to be determined on the basis of the effect of ionic interactions. In an aqueous solution, negative ions become surrounded by the cloud of positive ions, and vice versa. The real system would have lower Gibbs free energy than a hypothetical system where the ions are completely distributed randomly. In the real system, the electrostatic interaction generates an activity coefficient to be less than unity. In an ideal solution or infinite dilution system, an activity coefficient would be close to the unity. When the ions are point charges and the interactions are entirely electrostatic, the distribution of ions around any particular ion follows the Boltzmann distribution. Then, the activity coefficient is given by original Debye-Hückel model as defined in Eq. (3.12).

$$\log \gamma_i = -Az_i^2\sqrt{I} \quad (3.12)$$

where A is the temperature-dependent constant, z_i is the charge of the ion, and I is the ionic strength of the solution. The ionic strength takes account of electrostatic effectiveness of polyvalent ions. It is defined as a function of molality of ions and charge of ion as shown in Eq. (3.13).

$$I = \frac{1}{2}\sum m_i z_i^2 \tag{3.13}$$

The original Debye-Hückel model is reasonable up to ionic strength with 1×10^{-3} molal concentration. At the higher ionic strength, the original Debye-Hückel becomes less accurate to predict the activity coefficient. The extended Debye-Hückel model of Eq. (3.14) is applicable to the diluted electrolyte solution systems, which have ionic strength with less than 0.1.

$$\log \gamma_i = -\frac{A z_i^2 \sqrt{I}}{1 + B\dot{a}\sqrt{I}} \tag{3.14}$$

where B is the temperature-dependent constant and $\dot{a}_i$ is the empirical ion-size parameter measuring the effective diameter of the hydrated ion.

In addition, the Davies equation of Eq. (3.15) also calculates the activity coefficient. It is a variant of the extended Debye-Hückel model. It is capable to be used up to ionic strength up with 0.5 molal concentration.

$$\log \gamma_i = -A z_i^2 \left(\frac{\sqrt{I}}{1 + \sqrt{I}} - 0.3I\right) \tag{3.15}$$

Another modified version of the extended Debye-Hückel model is the B-dot model as defined in Eq. (3.16). It is parameterized from 0 to 300°C for the solutions of up to 3 molal ionic strength (Bethke, 1996).

$$\log \gamma_i = -\frac{A z_i^2 \sqrt{I}}{1 + B\dot{a}\sqrt{I}} + \dot{B}I \tag{3.16}$$

where $\dot{B}$ is the temperature-dependent constant.

The previous Debye-Hückel, Davies, and B-dot models depend on ionic strength and charge of ions. Debye-Hückel and B-dot models, additionally, are affected by the ion size and temperature-dependent constants. The activity coefficient calculated by Debye-Hückel model becomes unity when ionic strength decreases to zero. In the high ionic strength condition, the real activity coefficient increases, but calculated coefficient calculated by Debye-Hückel model still decreases. Therefore, above the 0.1 molal concentration of ionic strength, Debye-Hückel has less accuracy and other models are recommended to be used. The Davies model gives more reasonable activity coefficient in the range of 0.3–0.5 molal concentrations with ionic strength. Above the ionic strength of 0.5 molal concentration, it predicts the coefficient to increase. The B-dot model is known as to calculate the accurate activities of Na^+ and Cl^- up to several molal concentrations of ionic strength. It can reasonably estimate the activities of other species when ionic strength is up to 0.3–1 molal concentrations (Bethke, 1996).

Basic Reactions of Geochemistry

Aqueous reactions

In aqueous solutions, ions can be attached to other ions or complexes and form other complexes. Because these reactions in aqueous solutions are incorporated in one phase, they are homogeneous reactions. The formation of aqueous complexes is relatively fast reaction and follows the equilibria. The aqueous equilibrium reaction between the components in an aqueous phase follows the law of mass action. For an example of the aqueous reaction corresponding to Eq. (3.17), the distribution of the species is obtained by the equilibrium constant of Eq. (3.18). The equilibrium constant is also termed as a stability constant.

$$Ca^{2+} + SO_4^{2-} \leftrightarrow CaSO_4 \tag{3.17}$$

$$K_{eq} = \frac{[CaSO_4]}{[Ca^{2+}][SO_4^{2-}]} = 10^{2.5} \tag{3.18}$$

Mineral reactions

The mineral reactions of dissolution and precipitation are heterogeneous reactions because the species involved in these reactions are in the different phases, i.e., solid and aqueous phases. In the reaction of calcium carbonate corresponding to Eq. (3.19), the law of mass action describes the equilibrium state of mineral reactions. In the expression of equilibrium constant of mineral reactions, the activity of a pure solid is unity and the equilibrium constant is termed as the solubility product of Eq. (3.20).

$$CaCO_3 \leftrightarrow Ca^{2+} + CO_3^{2-} \tag{3.19}$$

$$K_{sp} = [Ca^{2+}][CO_3^{2-}] = 10^{-8.48} \text{ at } 25^\circ C \tag{3.20}$$

where the K_{sp} is the solubility product.

Generally, the mineral reactions are slow kinetic reactions, and the achievement of equilibrium of mineral reactions requires relatively longer time than that of aqueous reactions. A saturation index is defined to determine whether the mineral reactions are under equilibrium or not. The saturation index is defined as the logarithm of a saturation state of Eq. (3.21). The saturation state of Eq. (3.22) is defined as the ratio of ion activity product (IAP) to solubility product.

$$SI = \log \Omega \tag{3.21}$$

$$\Omega = \frac{IAP}{K_{sp}} \tag{3.22}$$

where SI indicates the saturation index, Ω is the saturation state, and IAP is the ion activity product. The IAP is defined as the product of the activities of species in the water, and the IAP of calcium carbonate is described in Eq. (3.23).

$$IAP = \left[Ca^{2+}\right]\left[CO_3^{2-}\right] \tag{3.23}$$

The saturation index is a useful tool to state the saturation condition of mineral reaction. When the saturation index is equal to zero, there is an equilibrium between the mineral and the solution. For the saturation index is less than zero, there is a subsaturation state resulting in mineral dissolution. For the saturation index is higher than zero, there is a supersaturation state resulting in mineral precipitation.

The dissolution or precipitation of minerals is the result of multiple processes: (1) the transport of solutes between solutions and the mineral surface, (2) the adsorption and desorption of solutes at the surface, (3) the hydration and dehydration of ions, and (4) surface migration. Some processes are faster than the other processes. The slowest process dominates the overall rate of the mineral reaction. The slow mineral reactions are considered as kinetically controlled reactions. There are a number of theories to describe the reaction kinetics. One of the successful theories is the general rate law of transition state theory (TST) proposed by Eyring (1935). The reaction rate of the mineral reaction is described in Eq. (3.24).

$$r = \widehat{A}k_a \exp\left(-\frac{E_a}{RT}\right)a_{H^+}\left(1-\frac{\Omega}{K_{sp}}\right) \tag{3.24}$$

where r is the reaction rate, $\widehat{A}$ is the reactive surface area of the mineral, k_a is the reaction rate constant, and E_a is the activation energy of the reaction. The reaction rate depends on the pH. In addition, the reaction rate constant is sensitive to the temperature following Eq. (3.25).

$$k_a = k_O \exp\left[-\frac{E_a}{R}\left(\frac{1}{T}-\frac{1}{T_O}\right)\right] \tag{3.25}$$

where k_O is the reaction rate constant at reference temperature and T_O is the reference temperature.

Ion exchange

Clay mineral exhibits an ion-exchange behavior with cations. The equilibrium of ion exchange between two cations corresponding to Eq. (3.26) follows Eq. (3.27). The equilibrium constant of ion exchange is termed with an exchange constant.

$$\text{A-X} + \text{B}^+ \leftrightarrow \text{B-X} + \text{A}^+ \tag{3.26}$$

$$K_{A/B} = \frac{[\text{A-X}][\text{B}^+]}{[\text{A}^+][\text{B-X}]} \tag{3.27}$$

where X represents the clay surface, A and B are cations, A-X and B-X are the adhered cations on the clay surface, and $K_{A/B}$ is the exchange constant.

Because the activity coefficients of A-X and B-X are applicable to the none of Debye-Hückel, Davies, and B-dot models, a selectivity coefficient is normally used rather than the exchange constant. In the definition of selectivity coefficient, the activities of A-X and B-X are replaced by equivalent fractions of A-X and B-X. The equivalent fractions are defined as the fractions of cation exchange capacity (CEC) occupied by the particular ion. Introducing the concepts of selectivity coefficient, equivalent fraction, and activity of ions, the equilibrium constant is rewritten as in Eq. (3.28). The equivalent fraction is also defined as in Eq. (3.29).

$$K'_{A/B} = \frac{\zeta(\text{A-X})m(\text{B}^+)\gamma(\text{B}^+)}{\zeta(\text{B-X})m(\text{A}^+)\gamma(\text{A}^+)} \tag{3.28}$$

$$\zeta(i\text{-X}) = \frac{meq_{i\text{-X}}}{CEC} = \frac{meq_{i\text{-X}}}{\sum_{A,B,\ldots} meq_{i\text{-X}}} \tag{3.29}$$

where $K'_{A/B}$ indicates the selectivity coefficient, $\zeta(\text{A-X})$ and $\zeta(\text{B-X})$ are the equivalent fractions of A-X and B-X, $meq_{i\text{-X}}$ is the milliequivalent of the exchangeable species i, and CEC is the cation exchange capacity.

There are three common conventions (Gaines-Thomas, Vanselow, and Gapon) to describe the selectivity coefficient of ion exchange. For the homovalent exchange reaction, there is no impact what convention is applied. For the heterovalent exchange reaction, the type of convention results in the different results of ion exchange. For the exchange between sodium and calcium cations, the Gaines-Thomas convention describes the ion-exchange reaction and selectivity coefficient as Eqs. (3.30) and (3.31).

$$\text{Na}^+ + \frac{1}{2}\text{Ca-X}_2 \leftrightarrow \text{Na-X} + \frac{1}{2}\text{Ca}^{2+} \tag{3.30}$$

$$K'_{Ca/Na} = \frac{\zeta(\text{Na-X})\left[\text{Ca}^{2+}\right]^{0.5}}{\zeta(\text{Ca-X}_2)^{0.5}\left[\text{Na}^+\right]} \tag{3.31}$$

In the descriptions of Eqs. (3.30) and (3.31), Vanselow convention uses the molar fraction instead of the equivalent fraction. The Gapon convention describes the activities of adhered ions with a fraction of the number of exchange sites as shown in Eq. (3.32).

The selectivity coefficient following Gapon convention is written in Eq. (3.33).

$$Na^{+} + Ca_{0.5}\text{-}X \leftrightarrow Na\text{-}X + \frac{1}{2}Ca^{2+} \tag{3.32}$$

$$K'_{Ca/Na} = \frac{\zeta(Na\text{-}X)\left[Ca^{2+}\right]^{0.5}}{\zeta(Ca_{0.5}\text{-}X)\left[Na^{+}\right]} \tag{3.33}$$

Surface complexations

In addition to the ion exchange, there are other types of the adsorption of the species to the surface of solid. The surface complexation describes the attachments of species to the existing functional groups of the solid surface of amorphous aluminosilicates, metal oxides/hydroxides, and organic matters (Al-Shalabi & Sepehrnoori, 2017). Generally, there are three models to describe the surface complexation: (1) constant capacitance model, (2) the diffuse layer model, and (3) the triple layer model. Practically, a double layer model can be applicable to comprehensive fluid-based systems and is composed of the constant capacitance model and diffuse layer model, generally, via the Gouy-Chapman model.

In the double layer model, the description of the surface complexation using the Gibbs free energy change has two terms. The first term is a chemical bond between the ions and surface atoms. The second one is an electrostatic effect affected by the surface charge. Therefore, the Gibbs free energy of surface complexation can be described in Eq. (3.34).

$$\Delta G_{total} = \Delta G_{chem} + \Delta G_{coul} \tag{3.34}$$

where ΔG_{total} is the total ΔG of surface complexation, ΔG_{chem} is the ΔG of the intrinsic chemical reactions at the surface, and ΔG_{coul} is the ΔG of electrostatic or Coulombic interaction.

The electrostatic term counts for the electrostatic interaction between an ion and a charged surface. It is the difference between energy state of 1 mole of ion at the surface and that in the bulk of the solution as defined in Eq. (3.35).

$$\Delta G_{coul} = \Delta G_{\psi=\psi_0} - \Delta G_{\psi=0} = z_i F(\psi_0 - 0) = z_i F \psi_0 \tag{3.35}$$

where $\Delta G_{\psi=0}$ is the ΔG of 1 mole ion in the bulk of the solution and equal to zero, $\Delta G_{\psi=\psi_0}$ indicates the ΔG of 1 mole ion at the surface, F is the Faraday constant, and ψ_0 is the surface potential.

Introducing Eq. (3.9), the Gibbs free energy change of the surface complexation model is rewritten with mass action constant as shown in Eqs. (3.36) and (3.37).

$$RT \ln K_a = -\Delta G_{total} = RT \ln K_{int} - z_i F \psi_0 \tag{3.36}$$

$$\ln K_a = \ln K_{int} - \frac{z_i F \psi_0}{RT} \tag{3.37}$$

where K_a is the apparent dissociation constant and K_{int} is the intrinsic dissociation constant.

The apparent dissociation constant varies with the surface potential of rock and can be measured experimentally. The intrinsic dissociation constant is a constant describing the chemical binding. As an example for the surface complexation, the surface protolysis reaction is shown in Eq. (3.38). Because the charge valence of proton is unity, the apparent dissociation constant of the surface protolysis reaction is depicted in Eqs. (3.39) and (3.40).

$$\equiv SOH + H^{+} \leftrightarrow \equiv SOH_2{}^{+} \tag{3.38}$$

$$K_a = \frac{\left[\equiv SOH_2{}^{+}\right]}{\left[\equiv SOH\right]\left[H^{+}\right]} \tag{3.39}$$

$$K_a = K_{int} \exp\left(-\frac{F\psi_0}{RT}\right) \tag{3.40}$$

where the prefix "≡" represents the species at the solid surface.

EMPIRICAL MODELING WITHOUT GEOCHEMISTRY

Research group of BP has published the numerical studies of LSWF modeling in sandstone reservoirs (Jerauld, Webb, Lin, & Seccombe, 2006, Jerauld, Webb, Lin, & Seccombe, 2008). They hypothesized the potential mechanism for increased oil recovery of LSWF is the wettability alteration. They modeled the wettability alteration of LSWF and, approximately, physical dispersion of mixing between connate water and injected water. Conventionally, the wettability modification is modeled through the modification of relative permeability and capillary pressure curves. The studies have adapted a simple empirical approach to model the wettability modification because of unclear predictive physics to explain the wettability modification. Experimental works in the previous chapters have demonstrated that the incremental oil recovery depends on the salinity of the brine, but it is not simply proportional to the salinity. Therefore, the empirical approach assumes that salinity-dependent residual oil saturation is an interpolation factor to modify relative permeability and capillary pressure. In the approach, the residual oil saturation linearly depends on the salinity between the low and high threshold salinities and is to be constant beyond the threshold conditions (Fig. 3.1). Therefore, the relative permeability and capillary pressure curves are linearly modified between

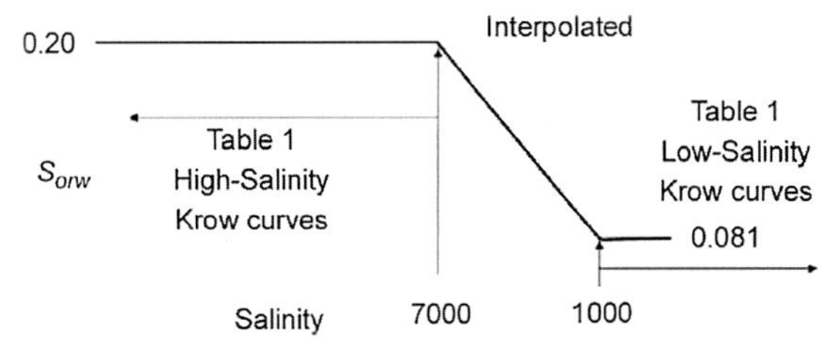

FIG. 3.1 The schematic description of salinity dependence of residual oil saturation used in the empirical approach. (From Jerauld, G. R., Webb, K. J., Lin, C.-Y., & Seccombe, J. C. (2008). Modeling low-salinity waterflooding. *SPE Reservoir Evaluation and Engineering, 11*(6), 1000–1012. https://doi.org/10.2118/102239-PA.)

low and high salinity threshold conditions using a normalized residual oil saturation. The salt is assumed to be an additional single-lumped component in an aqueous phase. The viscosity and density of the aqueous phase are the function of the salinity. Following Eqs. (3.41)–(3.45) formulate the empirical approach of wettability modification modeling.

$$k_{rw} = F_{IF}k_{rw}^{HS}(S_o^*) + (1 - F_{IF})k_{rw}^{LS}(S_o^*) \quad (3.41)$$

$$k_{ro} = F_{IF}k_{ro}^{HS}(S_o^*) + (1 - F_{IF})k_{ro}^{LS}(S_o^*) \quad (3.42)$$

$$p_c = F_{IF}p_c^{HS}(S_o^*) + (1 - F_{IF})p_c^{LS}(S_o^*) \quad (3.43)$$

$$F_{IF} = \frac{(S_{or} - S_{or}^{LS})}{(S_{or}^{HS} - S_{or}^{LS})} \quad (3.44)$$

$$S_o^* = \frac{(S_o - S_{or})}{(1 - S_{wi} - S_{or})} \quad (3.45)$$

where k_{rw} is the relative permeability of aqueous phase, F_{IF} is the interpolation factor, k_{rw}^{HS} is the relative permeability of aqueous phase at the high threshold salinity condition, S_o^* is the normalized residual oil saturation, k_{rw}^{LS} is the relative permeability of aqueous phase at the low threshold salinity condition, k_{ro} is the relative permeability of oleic phase, k_{ro}^{HS} is the relative permeability of oleic phase at the high salinity threshold condition, k_{ro}^{LS} is the relative permeability of oleic phase at the low salinity threshold condition, p_c is the capillary pressure, p_c^{HS} is the capillary pressure at the high salinity threshold condition, p_c^{LS} is the capillary pressure at the low salinity threshold condition, S_{or} is the residual oil saturation, S_{or}^{LS} is the residual oil saturation at the low salinity threshold condition, S_{or}^{HS} is the residual oil saturation at the high salinity threshold condition, and S_{wi} is the irreducible water saturation.

In addition, these studies cautioned the phenomena including connate water banking and physical dispersion for the simulations of LSWF process. The connate water banking will be formed when connate water and oil are displaced by low-salinity water. It flows ahead of the front of low-salinity water. The studies validated the effect of connate water banking on the displacement through extended Buckley-Leverett solution. It is explained that the higher oil recovery at breakthrough hardly guarantees the disappearance of the connate water banking. Therefore, it is concluded that the laboratory experiments of LSWF should consider the effect of connate water banking to interpret the salinity-dependent relative permeability curves to be used in simulations of LSWF. Because the mixing between connate water and low-salinity water influences the interpretation and predictions of LSWF, the physical dispersion is also of importance in both laboratory and field tests. In the numerical simulations of LSWF, the physical dispersion is approximated by using the numerical dispersion. Although the numerical dispersion fairly simulates the physical dispersion, the accurate modeling of the dispersion is sensitive to the grid size and dimension of numerical simulation. In addition, the level of dispersion depends on the slug size. Even though the simulation uses a fine grid model to capture the physical dispersion with the numerical dispersion, it requires significant computational time. Incorporating the pseudo-relative permeability and modified salinity dependency, the simulation with a coarse grid model can provide the same results of the fine grid model simulation and requires less simulation time. Using the empirical approach of wettability modification and modeling of numerical dispersion, they have simulated the LSWF process with the physical dispersion.

MECHANISTIC MODELING WITH GEOCHEMISTRY

Omekeh, Friis, Fjelde, and Evje (2012) developed a numerical model of Buckley-Leverett two-phase flow to simulate the core flooding of LSWF for sandstone reservoirs. This study suggested that the LSWF modifies the wettability of sandstone based on the MIE mechanism, not pH increase, and the pH increase is only the result of mineral dissolution and aqueous solubility of CO_2. The numerical simulation employs the ion-exchange reactions to model the MIE mechanism. Because the geochemistry is closely related to the LSWF process, the modeling of the Buckley-Leverett two-phase flow incorporates the modeling of geochemical reactions of aqueous reactions, mineral reactions, and the ion exchange. In addition, there is an assumption for the modeling of aqueous solubility of CO_2. The solubility of CO_2 in water is not calculated by thermodynamic modeling, and some CO_2 is assumed to be dissolved in water. With this assumption, the dissolved CO_2 in water participates in the aqueous and mineral reactions. In terms of wettability modification modeling, the study adopted an approach following the MIE mechanism, which is assumed to modify relative permeability through the ion-exchange reaction. The injection of low-salinity water results in the detachment of the divalent cations (Ca^{2+} and Mg^{2+}) because of the ion exchange. The amount of the detachment linearly interpolates the relative permeability between threshold high and low salinity conditions as shown in Eqs. (3.41) and (3.42). The interpolation factor is given by Eq. (3.46) and a function of concentration of the adhered divalent cation as shown in Eq. (3.47).

$$F_{IF} = F_{IF}(\beta_{Ca}, \beta_{Mg}) = \frac{1}{1 + am(\beta_{Ca}, \beta_{Mg})} \tag{3.46}$$

$$m(\beta_{Ca}, \beta_{Mg}) = \max[\beta_{Ca}(t=0) - \beta_{Ca}(t>0), 0] + \max[\beta_{Mg}(t=0) - \beta_{Mg}(t>0), 0] \tag{3.47}$$

where β_{Ca} and β_{Mg} indicate the desorbed Ca^{2+} and Mg^{2+}, a is the constant, t is the time, and $m(\beta_{Ca}, \beta_{Mg})$ is the amount of the desorbed Ca^{2+} and Mg^{2+}.

For validating the LSWF model, the study numerically simulated the experimental corefiooding of LSWF by Fjelde, Asen, and Omekeh (2012). Both experiment and numerical simulation successively injected formation water, seawater, and low-salinity water. They compared the effluent concentrations of the divalent cations (Ca^{2+} and Mg^{2+}) (Fig. 3.2). In the experiment, the retention in the effluent concentration of Ca^{2+} is

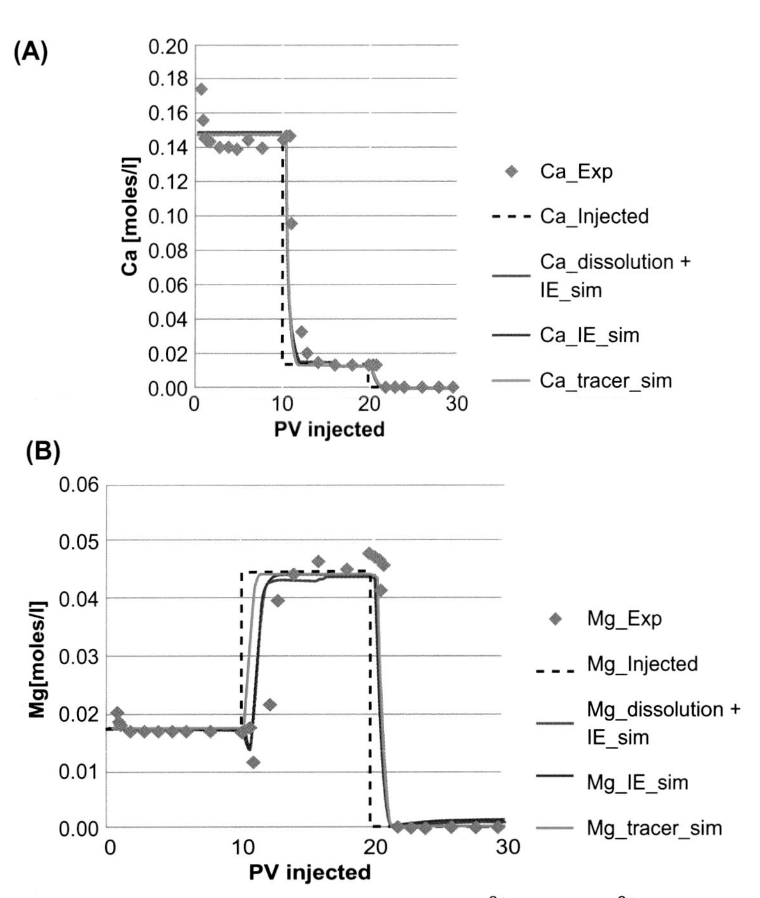

FIG. 3.2 The comparison of the effluent concentrations of **(A)** Ca^{2+} and **(B)** Mg^{2+} between simulations and experiments. (From Fjelde, I., Asen, S. M., & Omekeh, A. V. (2012). Low salinity water flooding experiments and interpretation by simulations. In: *Paper presented at the SPE improved oil recovery symposium, Tulsa, Oklahoma, USA, 14–18 April*. https://doi.org/10.2118/154142-MS.)

observed when injecting fluid is switched from seawater to low-salinity water. The numerical simulation indicates that the ion exchange causes the high retention of Ca^{2+}, but carbonate mineral dissolution generating Ca^{2+} in water slightly reduces the degree of retention. The combined geochemical reactions contribute to the retention of effluent history of Ca^{2+}. In the experiment, a slight reduction in the effluent concentration of Mg^{2+} is also observed when the injecting fluid is switched from formation water to seawater. The numerical simulation explains that ion-exchange reaction reduces the effluent concentration of Mg^{2+}. Because the seawater has higher concentration of Mg^{2+} over formation water, the ion-exchange reaction replaces the adhered ions with Mg^{2+} on the negative sites of the clay surface. The study also examined the effluent pH as the result of geochemical reactions. In the results of pH, the numerical simulation illustrates that the dissolution of calcite is sensitive to the pH of brine. The low-salinity water has relatively high pH, and the injection of low-salinity water contributes to less mineral dissolution compared with the injections of formation water and seawater. This numerical simulation study successfully developed the LSWF process coupled with geochemistry and explained the experimental results from a geochemical point of view. However, the numerical model employs the limited number of geochemical reactions.

Nghiem, Sammon, Grabenstetter, and Ohkuma (2004) advanced the GEM software, developed by CMG (Computer Modelling Group, Ltd.), coupled with the comprehensive geochemical reactions. The GEM software is the multiphase, multicomponent, and equation of state (EOS) simulator and also has a capability to model the geochemical reactions. Later, Dang, Nghiem, Chen and Nguyen (2013) advanced the GEM simulator to model the LSWF considering comprehensive geochemical reactions. Before the description of work of Dang et al. (2013), the important features of Nghiem et al. (2004) are discussed.

The purpose of the Nghiem et al. (2004) is to model the CO_2 storage process in saline aquifers. The geochemical reactions include the aqueous reaction, mineral dissolution and precipitation, and ion exchange. The gaseous CO_2 has the solubility in the aqueous phase, and the aqueous solubility of CO_2 is of importance in the mechanisms of CO_2 storage. Because the dissolved CO_2 in brine influences the pH of brine and mineral reactions are highly affected by the pH of brine, the accurate modeling of CO_2 dissolution in the aqueous phase is required. Although the numerical study of Omekeh et al. (2012) considered the effect of solubilized CO_2 in aqueous phase on the geochemistry, it assumed the aqueous solubility without thermodynamics calculations. However, Nghiem et al. (2004) calculated the aqueous solubility of CO_2 incorporating the equilibrium relationship between aqueous and gases phases. The equilibrium relation of CO_2 between aqueous and gaseous phases is described in Eq. (3.48). A phase equilibrium process determines the solubility of CO_2 in aqueous phase and is modeled by the equality of fugacities between aqueous and gaseous phases as shown in Eq. (3.49). The EOS model (Peng & Robinson, 1976) calculates the fugacity in gaseous phase, and either Henry's law (Li & Nghiem, 1986) or Søreide-Whitson-Peng-Robinson (Søreide & Whitson, 1992) calculates the fugacity in aqueous phase. Eq. (3.50) indicates the Henry's law, and the Henry's constant in the law is a function of temperature and pressure following Eq. (3.51).

$$CO_2(aq) \leftrightarrow CO_2(g) \tag{3.48}$$

$$f_{i,g} = f_{i,aq} \tag{3.49}$$

$$f_{i,aq} = H_i x_i \tag{3.50}$$

$$\ln H_i = \ln H_i^s + \frac{1}{RT}\int_{p_{H_2O}^s}^{p} \overline{v}_i dp \tag{3.51}$$

where $f_{i,j}$ indicates the fugacity of species i, i.e., CO_2, in the phase j, H_i is the Henry's constant of species i, x_i is the mole fraction of species i in aqueous phase, H_i^s is the Henry's constant at the saturation pressure of H_2O, temperature, and zero salinity, $p_{H_2O}^s$ is the saturation pressure of H_2O, and $\overline{v}_i$ is the partial molar volume of species i.

The Henry's constant at the H_2O saturation pressure and temperature also depends on the pressure and temperature, and it is determined with the following Harvey's relation (Harvey, 1996). Harvey (1996) published the correlations of Henry's constant at the H_2O saturation pressure and temperature to handle the effects of pressure and temperature for a few gaseous components (CO_2, N_2, H_2S, and CH_4). For CO_2, the following correlation of Eq. (3.52) estimates pressure and temperature dependent of the Henry's constant on the basis of saturation pressure of H_2O and reduced temperature. Saul and Wagner (1987) provided the saturation pressure of H_2O at the temperature. The partial molar volume of CO_2 is also defined to calculate the Henry's constant. Garcia (2001) presented the correlation of partial molar volume of CO_2 as a function of temperature as shown in Eq. (3.53).

$$\ln H^s_{CO_2} = \ln p^s_{H_2O} + -9.4234(T_{r,H_2O})^{-1} + 4.0087(1 - T_{r,H_2O})^{0.355}(T_{r,H_2O})^{-1} + 10.3199[\exp(1 - T_{r,H_2O})](T_{r,H_2O})^{-0.41} \quad (3.52)$$

$$\bar{v} = 37.51 - 9.585 \times 10^{-2}T + 8.740 \times 10^{-4}T^2 - 5.044 \times 10^{-7}T^3 \quad (3.53)$$

where T_{r,H_2O} is the reduced temperature of H_2O and T is the temperature (°C).

In addition to the pressure and temperature, the salinity of the system also influences the aqueous solubility of CO_2. The Henry's constant varies according to the salinity of brine. It is reported that the solubility of the gaseous components generally decreases with increasing salinity. It is known as the salting-out phenomenon. With the calculated Henry's constant at zero salinity, following relation of Eq. (3.54) takes salting-out phenomenon into consideration. The salting-out coefficient in the equation is sensitive to the temperature, and Bakker (2003) formulated the correlations between the salting-out coefficient and temperature for the CO_2 and CH_4. For CO_2, Eq. (3.55) presents the correlation of salting-out coefficient of CO_2.

$$\ln\left(\frac{H_{salt,i}}{H_i}\right) = k_{salt,i}m_{salt} \quad (3.54)$$

$$k_{salt,CO_2} = 0.11572 - 6.0293 \times 10^{-4}T + 3.5817 \times 10^{-6}T^2 - 3.7772 \times 10^{-9}T^3 \quad (3.55)$$

where $H_{salt,i}$ is the Henry's constant of saline water of species i, $k_{salt,i}$ is the salting-out coefficient of species i, m_{salt} is the molality of the dissolved salt, and T is the temperature (°C).

Dang et al. (2013) advanced the GEM simulator of multicomponent and multiphase transport, EOS, and geochemistry to model LSWF process for sandstone reservoirs. The proposed numerical model using the simulator incorporates the important physical and chemical phenomena of LSWF. Firstly, this study validated the geochemical reaction modeling by comparing with the PHREEQC software (Parkhurst & Appelo, 1999). The PHREEQC is the comprehensive geochemistry software and enables to calculate (1) the speciation and saturation index; (2) the batch reaction, one-dimensional flow simulation with aqueous, mineral, gas, solid solution, surface complexation, and ion-exchange equilibria, irreversible reactions of specified mole transfers of reactants, kinetically controlled reactions, mixing of solutions, and temperature changes; and (3) inverse modeling. Before the simulation of EOR process, the one-dimensional flow simulation considering multicomponent, single-phase, and geochemical reactions is carried out. The study simulated the injection of low-salinity water into connate water-saturated core system using both simulators of GEM and PHREEQC. It compared the effluent concentrations of Ca^{2+} and Na^+ between the GEM and PHREEQC and observed the good match from the comparisons. Secondly, the study tested and validated the numerical model of LSWF by comparing with the experimental results (Fjelde et al., 2012; Rivet, 2009). In the study, numerical simulation agrees the wettability modification because of ion exchange as the main mechanism behind the LSWF. It assumes and models the wettability modification of LSWF as a function of the equivalent fraction of cation. It linearly interpolates both the oil and water relative permeabilities between high and low threshold salinity conditions according to the equivalent fraction of Ca^{2+} (Fig. 3.3). Following the approach of wettability modification modeling, the LSWF model simulates the two experiments of LSWF (Fjelde et al., 2012; Rivet, 2009). The effluent concentrations of Ca^{2+}, pH, and oil saturation are compared between the simulations and experiments to validate the feasibility of numerical

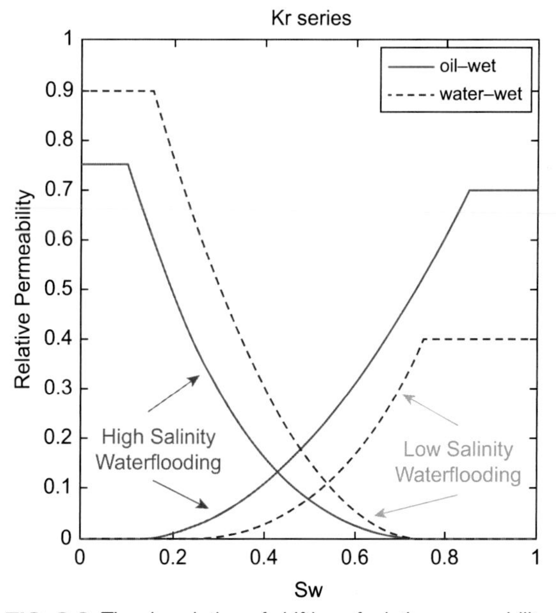

FIG. 3.3 The description of shifting of relative permeability curves between high and low threshold conditions. (From Dang, C. T. Q., Nghiem, L. X., Chen, Z. J., & Nguyen, Q. P. (2013). Modeling low salinity waterflooding: Ion exchange, geochemistry and wettability alteration. In: *Paper presented at the SPE annual technical conference and exhibition, New Orleans, Louisiana, USA, 30 September–2 October*. https://doi.org/10.2118/166447-MS.)

models of LSWF. For the experimental results of Fjelde et al. (2012), numerical simulations with and without cation-exchange reaction of Ca^{2+} are investigated in terms of the effluent concentration of Ca^{2+}. The core flooding process is designed to inject seawater after formation water and then low-salinity water after the seawater. The simulation without the cation exchange shows a discrepancy to the experimental result during seawater injection, but the simulation with the cation exchange provides a great match (Fig. 3.4A). The simulation with the cation-exchange reaction also shows an excellent match of the effluent pH against the experimental measurement (Fig. 3.4B). The numerical simulation describes an increase in the effluent pH, as the injecting brine salinity decreases. The trend of increasing pH can be explained with the dissolution calcite mineral during LSWF. The dissolution of calcite mineral by LSWF consumes the hydrogen ion and increases the pH. In addition, the modeling of wettability modification sufficiently captures the reduction in remained oil saturation (Fig. 3.4C). For the experiment (Rivet, 2009), the LSWF simulation with ion-exchange reaction also successfully describes the historical results of residual oil saturation and effluent pH.

Kazemi Nia Korrani, Sepehrnoori, and Delshad (2013) from the University of Texas at Austin advanced the UTCHEM, an in-house chemical flood simulator, by coupling with IPhreeqc, a geochemical module by the United States Geological Survey (USGS). The IPhreeqc is the open source module of the PHREEQC software. The advanced UTCHEM can be applied to various EOR processes, including alkali and surfactant floods in both sandstone and carbonate reservoirs. It also has a capability to mechanistically model LSWF introducing comprehensive geochemistry. This simulator models the wettability modification of LSWF as a function of geochemical ions and reactions. A series of studies (Al-Shalabi, Luo, Delshad, & Sepehrnoori, 2015; Al-Shalabi, Sepehrnoori, Delshad, & Pope, 2015; Al-Shalabi, Sepehrnoori, Pope, & Mohanty, 2014) have used the UTCHEM to construct the numerical model of LSWF process in carbonate reservoirs. They simulated the coreflooding experiments of carbonate (Mohanty & Chandrasekhar, 2013; Yousef, Al-Saleh, and Al-Jawfi 2012; Yousef, Al-Saleh, Al-Kaabi, & Al-Jawfi, 2011) and reported the three methodologies of wettability modification modeling: (1) empirical model using the contact angle (Al-Shalabi, Sepehrnoori, et al., 2015); (2) fundamental model using the trapping number (Al-Shalabi et al., 2014); and (3) mechanistic model using the molar Gibbs free energy of solution (Al-Shalabi, Sepehrnoori, & Pope, 2015).

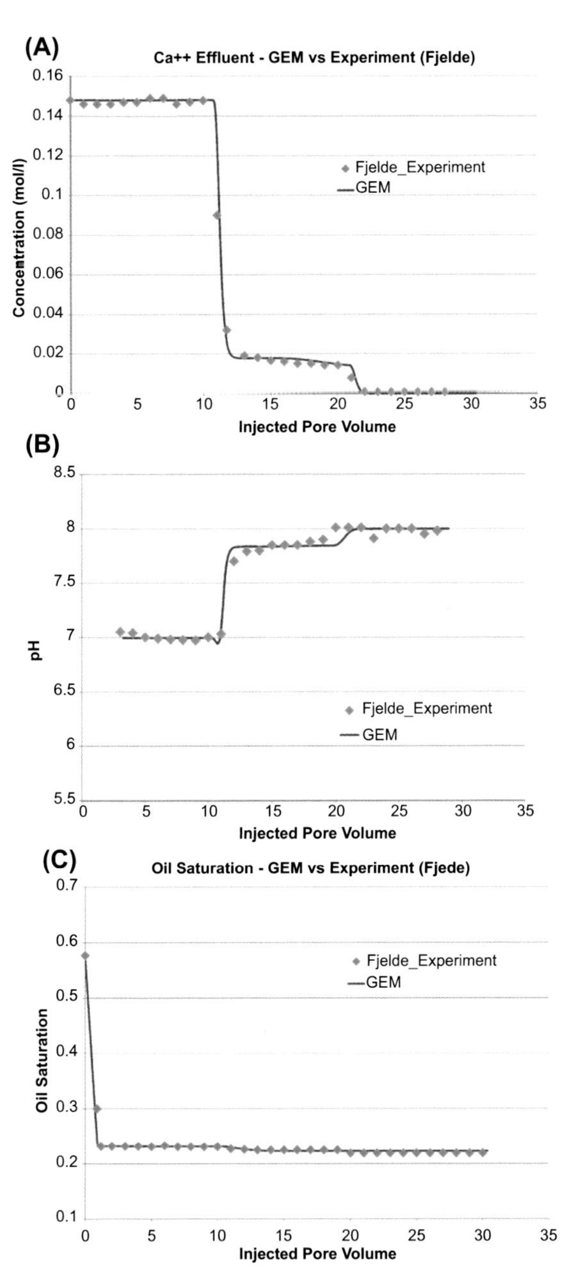

FIG. 3.4 The comparison of LSWF between simulation and experiment (Fjelde et al., 2012) in terms of the **(A)** effluent concentration of Ca^{2+}, **(B)** pH, and **(C)** remained oil saturation. (From Dang, C. T. Q., Nghiem, L. X., Chen, Z. J., & Nguyen, Q. P. (2013). Modeling low salinity waterflooding: Ion exchange, geochemistry and wettability alteration. In: *Paper presented at the SPE annual technical conference and exhibition, New Orleans, Louisiana, USA, 30 September–2 October*. https://doi.org/10.2118/166447-MS.)

The first methodology, empirical model, adopts the contact angle as an interpolation factor for wettability modification modeling. In addition, it introduces a third-degree polynomial relationship between contact angle and salinity. Using the polynomial relationship between contact angle and salinity, the residual oil saturation and oil relative permeability between low and high threshold conditions are interpolated by the salinity-dependent contact angle. The linear modification of residual oil saturation follows Eq. (3.56). The normalized contact angle of Eq. (3.57) rather than the contact angle is introduced to modify the residual oil saturation. The modification of oil relative permeability employs the modifications of endpoint and Corey's exponent of oil relative permeability. The interpolations of oil endpoint and Corey's exponent are functions of contact angle as shown in Eqs. (3.58) and (3.59). The validation of the empirical model is performed against the coreflooding of experiments. The proposed empirical model successfully matches the oil recoveries from the experiments.

$$S_{or} = \theta^* S_{or}^{LS} + (1 - \theta^*) S_{or}^{HS} \tag{3.56}$$

$$\theta^* = \frac{\theta - \theta^{HS}}{\theta^{LS} - \theta^{HS}} \tag{3.57}$$

$$k_{ro}^o = \frac{k_{ro}^{o\ LS} - k_{ro}^{o\ HS}}{1 + \left(\frac{\theta}{a}\right)^e} + k_{ro}^{o\ HS} \tag{3.58}$$

$$n_o = \frac{n_{o,\max} - n_o^{LS}}{1 + \left(\frac{\theta}{a}\right)^{-e}} + n_o^{LS} \tag{3.59}$$

where θ is the contact angle, θ^* is the normalized contact angle between high and low salinity threshold conditions, θ^{HS} and θ^{LS} are contact angle at the high and low threshold conditions, k_{ro}^o is the endpoint of oil relative permeability, $k_{ro}^{o\ HS}$ and $k_{ro}^{o\ LS}$ are the endpoints of oil relative permeability, respectively, at high and low threshold conditions, a is the inflection point from curve fitting, and e is the hill slope.

The second approach of fundamental model interpolates the residual oil saturation and oil relative permeability by the trapping number. For the horizontal system, the trapping number is defined with capillary number, the ratio of viscous to capillary forces, and Bond number, the ratio of gravity to capillary forces as shown in Eqs. (3.60)–(3.62).

$$N_T = \sqrt{N_c^2 + N_B^2} \tag{3.60}$$

$$N_c = \frac{v\mu}{\sigma \cos \theta} \tag{3.61}$$

$$N_B = \frac{kg\Delta\rho}{\sigma \cos \theta} \tag{3.62}$$

where N_T is the trapping number, N_c is the capillary number, v is the velocity, N_B is the Bond number, σ is the interfacial tension, k is the permeability, $\Delta\rho$ is the difference of densities between displacing and displaced fluids, and g is the gravitational acceleration.

In terms of the residual oil saturation, the fundamental model employs the modified version of capillary desaturation curve (CDC) (Pope et al., 2000) to relate the residual oil saturation with trapping number as shown in Eq. (3.63). For the endpoint and Corey's exponent modifications of oil relative permeability, it uses the linear and natural logarithm relations, which are functions of the trapping parameters and trapping number. Because the trapping number is a function of contact angle, as shown in Eqs. (3.60)–(3.62), the alteration of contact angle during LSWF modifies the residual oil saturation, endpoint, and Corey's exponent of oil relative permeability.

$$S_{or} = S_{or}^{high} + \frac{S_{or}^{low} - S_{or}^{high}}{1 + T_o (N_T)^{\tau_o}} \tag{3.63}$$

where S_{or}^{high} and S_{or}^{low} are the residual oil saturations at the high and low trapping numbers, T_o is the first trapping parameter, and τ_o is the second trapping parameter, which incorporates the effects of heterogeneity and initial oil saturation on residual oil saturation.

In the fundamental model, the LSWF is assumed to modify contact angle contributing to the trapping number. It employs two approaches modifying contact angle during LSWF. The first approach adopts the polynomial relation between contact angle and salinity, used in the first methodology of empirical model. The second approach for calculating contact angle considers the EDL thickness or Debye length as shown in Eq. (3.64). The Debye length is approximately determined by Eq. (3.65). The fundamental model successfully matches the historical observations of coreflooding experiments.

$$\theta = A + \frac{B}{\kappa^{-1}} \tag{3.64}$$

$$\kappa^{-1} = \sqrt{\frac{\varepsilon_r \varepsilon_0 k_B T}{2 N_A e^2 I}} \tag{3.65}$$

where A and B are the fitting parameters, N_A is the Avogadro constant, and e is the elementary charge.

The third approach of wettability modification modeling is the mechanistic model using the effective molar Gibbs free energy of solution. The effective molar Gibbs free energy of solution is defined as the summation of chemical potential of the all aqueous species (Eq. 3.66). In this mechanistic model, there are two approaches relating the modification of residual oil saturation and oil relative permeability to the effective molar Gibbs free energy of solution. The first approach of the mechanistic model incorporates the relationship between the molar Gibbs free energy of solution and contact angle. It applies the relationship to either of the previously proposed empirical or fundamental models correlating the contact angle to the modification of residual oil saturation and oil relative permeability. In the second approach, the residual oil saturation, oil endpoint, and oil Corey's exponent are direct functions of effective molar Gibbs free energy of solution. The linear relation is used to modify residual oil saturation by the effective molar Gibbs free energy, and the interpolation factor is a function of Gibbs free energy of solution as shown in Eqs. (3.67) and (3.68). The linear correlation also applies to the calculation of oil endpoint and oil Corey's exponent as a function of the effective molar Gibbs free energy. This mechanistic methodology is applied to the third coreflood of Yousef, Al-Saleh, et al. (2012) and fourth coreflood of Mohanty and Chandrasekhar (2013). It calculates the effective molar Gibbs free energy for both corefloods (Fig. 3.5). The value of the effective molar

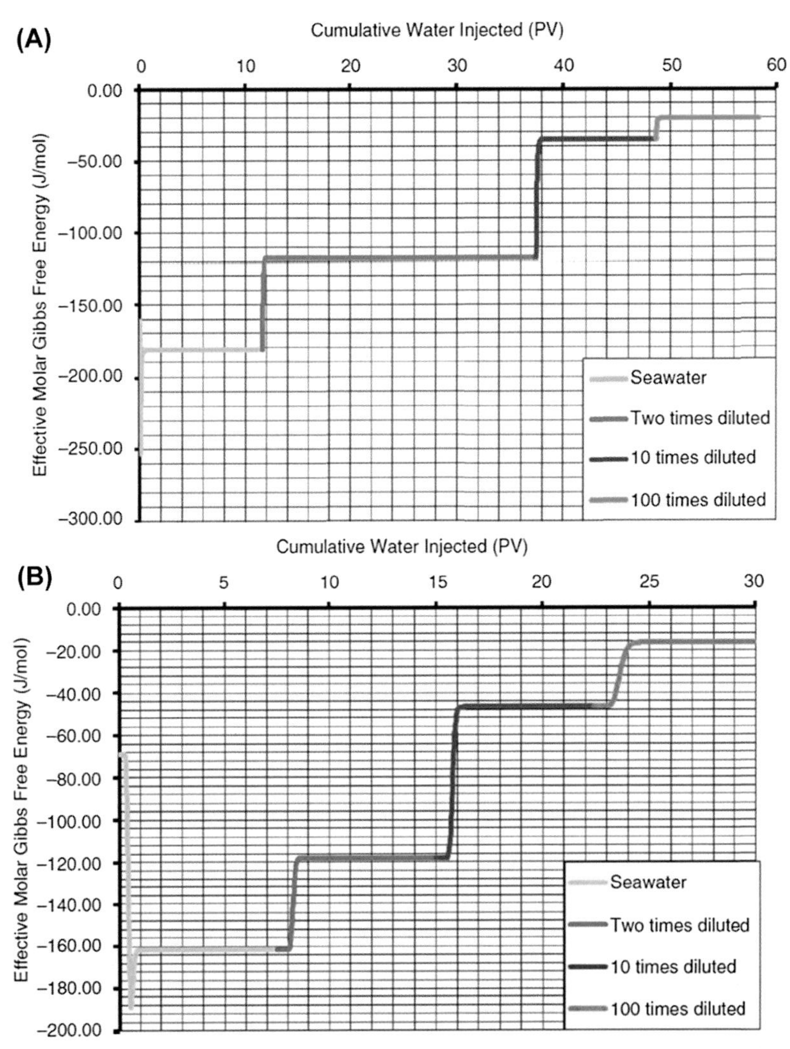

FIG. 3.5 The calculated effective molar Gibbs free energy of coreflooding experiments using the mechanistic model. (From Al-Shalabi, E. W., Sepehrnoori, K., & Pope, G. (2015). Mechanistic modeling of oil recovery due to low salinity water injection in oil reservoirs. In: *Paper presented at the SPE middle east oil & gas show and conference, Manama, Bahrain, 8–11 March*. https://doi.org/10.2118/172770-MS.)

Gibbs free energy decreases, as the salinity of injecting brine decreases. Maintaining the constant water relative permeability curve, the residual oil saturation, oil endpoint, and oil Corey's exponent are historically adjusted to reproduce the oil recovery and pressure drop of the coreflooding experiments.

$$\overline{G} = \sum_{i=1}^{N_{aq}} x_i \mu_i \quad (3.66)$$

$$S_{or} = F_{IF} S_{or}^{LS} + (1 - F_{IF}) S_{or}^{HS} \quad (3.67)$$

$$F_{IF} = \frac{\overline{G} - \overline{G}^{HS}}{\overline{G}^{LS} - \overline{G}^{HS}} \quad (3.68)$$

where $\overline{G}$ is the effective molar Gibbs free energy of solution, N_{aq} is the number of aqueous species i, and $\overline{G}^{HS}$ and $\overline{G}^{LS}$ are the effective molar Gibbs free energy of solution at high and low salinity threshold conditions.

The UTCOMP is another in-house simulator of the University of Texas at Austin and it is the EOS and compositional simulator to model the miscible and immiscible gas injection EOR processes. Kazemi Nia Korrani, Jerauld, and Sepehrnoori (2016) also advanced the UTCOMP coupled with the IPhreeqc module. In the coupling of UTCOMP and IPhreeqc, two additional reactions involving water-soluble hydrocarbon components and acidic/basic components of the hydrocarbon phase are implemented. Firstly, there are soluble hydrocarbon component and CO_2 in water. Especially, the dissolution of CO_2 in water influences the aqueous and mineral reactions. The dissolved CO_2 in water controls the pH owing to aqueous reactions, and the pH of brine affects the mineral reactions. The previous study of Nghiem et al. (2004) already implemented these reactions in the GEM software incorporating the equilibrium of fugacities between aqueous and gaseous phases. Kazemi Nia Korrani et al. (2016) also modeled the reactions of the water-soluble hydrocarbon component and CO_2 as well as the relevant geochemical reactions. Secondly, the acidic/basic components in hydrocarbon phase can be distributed between the aqueous and oleic phases (Havre, Sjöblom, & Vindstad, 2003). Following the equilibrium relationship of Eq. (3.69), the distribution of partitioned acids in each phase is determined by partition coefficient of Eq. (3.70). The partitioned acid in the aqueous phase dissociates in the aqueous phase as shown in Eqs. (3.71) and (3.72). The dissociation of the partitioned acid is also affected by the pH of brine.

$$HA_w \leftrightarrow HA_o \quad (3.69)$$

$$K_D = \frac{[HA_w]}{[HA_o]} \quad (3.70)$$

$$HA_w \leftrightarrow A_w^- + H^+ \quad (3.71)$$

$$K_a = \frac{[H^+][A_w^-]}{HA_w} \quad (3.72)$$

where HA_w and HA_o are the acidic component in the aqueous and oleic phases, A_w and A_o are the partitioned carboxylic organic components in aqueous and oleic phases, respectively, K_D is the partition coefficient, and K_a is the dissociation constant of HA_w in aqueous phase.

In addition to the modeling of comprehensive reactions, the numerical simulation study using the IPreeqc-UTCOMP modeled the LSWF process following wettability modification mechanism. The wettability modification is assumed to be controlled by a total ionic strength. The hypothetic simulations investigate the effects of the two additional reactions involving water-soluble hydrocarbon components and acidic/basic components of the hydrocarbon phase on the performance of LSWF process. The numerical simulations with and without the reactions clearly show the difference in the oil recovery (Fig. 3.6). Because the dissolution of CO_2 in aqueous phase potentially changes the total ionic strength, the simulations with and without the reactions have different degree of wettability modification and produce unequal oil recovery. Consequently, the study validated the numerical model of LSWF process through the simulations of the two coreflooding experiments (Kozaki, 2012). Using the reported parameters from Kozaki (2012), the coreflooding simulation of LSWF carries out the history matching process tuning unknown parameters. The history-matched model accurately reproduces the oil recovery as well as the effluent ion concentrations. The study also suggested the other approach to model wettability modification of LSWF process. The exchange reaction of organometallic complex (carboxylic organic component-divalent cation) by pure cations on the rock, potentially, changes the wettability of rock from oil-wet to water-wet (Fig. 3.7). Simplifying the reaction, the detachment of organometallic complex from the rock surface can be used as the interpolation factor for wettability modification modeling. The study briefly explained this approach and performed the hypothetical LSWF simulations with the approach of wettability modification modeling.

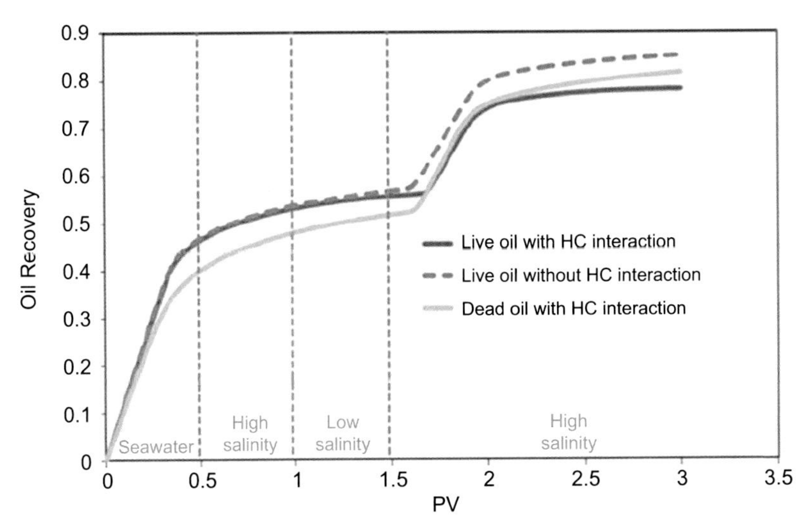

FIG. 3.6 The numerical simulations of low salinity waterflood with and without the reactions involving water-soluble hydrocarbon components and acidic/basic components of the hydrocarbon phase. (From Kazemi Nia Korrani, A., Jerauld, G. R., & Sepehrnoori, K. (2016). Mechanistic modeling of low-salinity waterflooding through coupling a geochemical package with a compositional reservoir simulator. *SPE Reservoir Evaluation and Engineering, 19*(1), 142–162. https://doi.org/10.2118/169115-PA.)

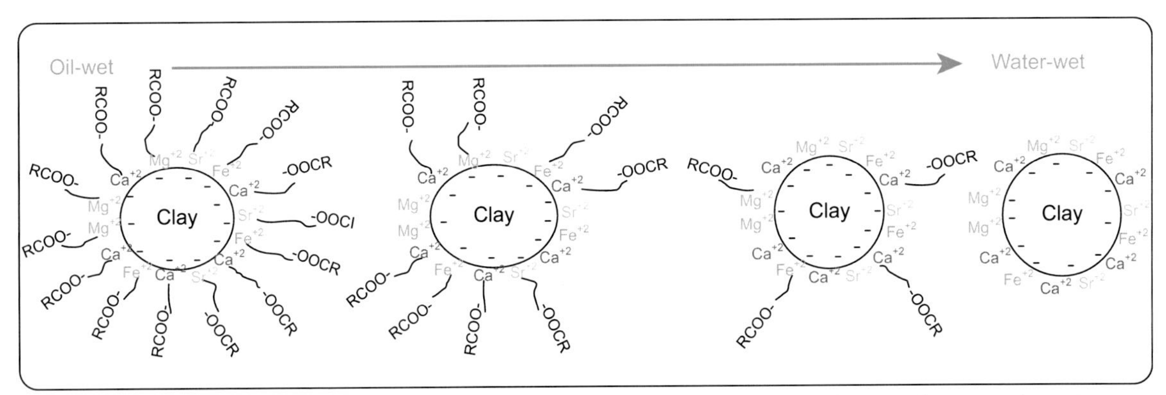

FIG. 3.7 The schematic description of relationship between wettability of a sandstone rock and organometallic complexes on the rock surface. (From Kazemi Nia Korrani, A., Jerauld, G. R., & Sepehrnoori, K. (2016). Mechanistic modeling of low-salinity waterflooding through coupling a geochemical package with a compositional reservoir simulator. *SPE Reservoir Evaluation and Engineering, 19*(1), 142–162. https://doi.org/10.2118/169115-PA.)

A series of studies (Brady & Krumhansl, 2012; Brady, Krumhansl, & Mariner, 2012; Brady et al., 2015; Brady and Thyne 2016) have proposed the surface complexation model to simulate the observations of LSWF processes in both sandstone and carbonate reservoirs. The studies (Brady & Krumhansl, 2012; Brady et al., 2012; Brady et al., 2015) have proposed the surface complexation model to describe the oil-water interface charge, clay edge surface charge, the adsorption of Ca^{2+} and Mg^{2+} on the clay edges, and reservoir-oil surface speciation (Table 3.1). Recalling Eq. (3.37), the surface complexation model is incorporated with a diffuse layer model to account for the electrical double layer effect. In the oil-water interface, the surface complexation model describes the two main reactions: (1) deprotonation of nitrogen and carboxylate groups of oil and (2) calcium-carboxylate surface complexation reaction. For the kaolinite and quartz minerals, there are the reactions of protonation/deprotonation. The adsorption of

TABLE 3.1
Surface Complexation Model in Sandstone Reservoir (Brady et al. 2015)

		Reactions
Oil surface		$-NH^+ \rightleftarrows -N + H^+$
		$-COOH \rightleftarrows -COO^- + H^+$
		$-COOH + Ca^{2+} \rightleftarrows -COOCa^+ + H^+$
Mineral surface	Quartz	$>SiOH \rightleftarrows >SiO^- + H^+$
		$>SiOH + Ca^{2+} \rightleftarrows >SiOCa^+ + H^+$
		$>SiOH + CaOH^+ \rightleftarrows >SiOCaOH + H^+$
	Kaolinite	$>AlOH_2{}^+ \rightleftarrows > AlOH + H^+$
		$>AlOH \rightleftarrows >AlO^- + H^+$
		$>SiOH \rightleftarrows >SiO^- + H^+$
		$>AlOH \rightleftarrows >AlO^- + H^+$
		$>SiOH + Ca^{2+} \rightleftarrows >SiOCa^+ + H^+$
		$>AlOH + Ca^{2+} \rightleftarrows >AlOCa^+ + H^+$
		$>SiOH + CaOH^+ \rightleftarrows >SiOCaOH + H^+$
		$>AlOH + CaOH^+ \rightleftarrows >AlOCaOH + H^+$
Kaolinite basal plane		$>H + Na^+ \rightleftarrows >Na + H^+$
		$2 > Na + Ca^{2+} \rightleftarrows >Ca + 2Na^+$
Oil-kaolinite		$>Al{:}Si{-}O^- + {}^+HN{-} \rightleftarrows >Al{:}Si{-}OHN{-}$
		$>Al{:}Si{-}O^- + {}^+CaOOC{-} \rightleftarrows >Al{:}Si{-}O{-}CaOOC$

Ca^{2+} and Mg^{2+} is accomplished by surface complexation model. Because of impure natural kaolinite, cation exchange occurs. Lastly, the surface complexation model of oil-reservoir surface speciation is used for the modeling of LSWF effect. The complexation model describes the adherences of the protonated nitrogen bases and positively charged calcium carboxylate groups to the negatively charged kaolinite edge. The surface complexation model describes the interaction of the oil-water-kaolinite system. It includes the submodels of (1) oil-water interface charge, (2) kaolinite edge surface charge, and (3) Ca^{2+} and Mg^{2+} sorption to oil and kaolinite edges. The diffuse layer model, which is the simplest model to describe the electric double layer, is used to develop the submodels of surface complexation. Using the surface complexation model, the concentration of species and electrostatic attraction concentration product can be determined at the equilibrium state (Fig. 3.8). The developed model is sensitive to ionic strength, temperature, and pH conditions. Brady et al. (2012) additionally proposed the surface complexation model of calcite minerals in sandstone reservoirs. Brady et al. (2015) introduced the concept of bond product sum (BPS) to indicate the mutual electrostatic adhesion, considering both the surface charge calculations of oil and kaolinite edges. The BPS is defined as the total of the products of the surface concentrations of oppositely charged species on the oil and minerals. When only negatively charged species exist on both the oil and mineral surfaces, the BPS is equal to zero because no oppositely charged, i.e., positively charged, species exists. There is no electrostatic adhesion meaning water-wetness. When only oil has positively charged surface species and mineral has the negatively charged surface species, the BPS is high enough to introduce the potential of adhesion, i.e., oil-wetness. The BPS is the indicator to imply the decreasing or increasing oil adhesion. The degree of BPS is controlled by the AN/BN of oil and pH. The isotherm disjoining pressure of oil and kaolinite edges based on the DLVO theory is calculated to complement the BPS estimate (Fig. 3.9). Brady and Thyne (2016) modeled the LSWF in the dolomite and limestone reservoirs by adapting the approach of surface complexation modeling (Fig. 3.10). Referring the concept of fractional wettability, potential linkage

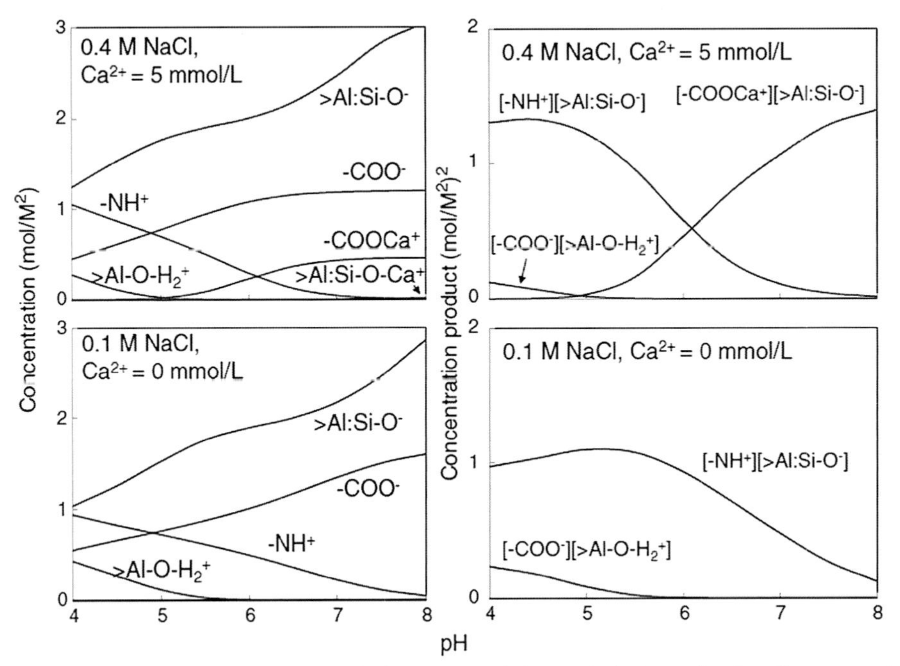

FIG. 3.8 Calculation of kaolinite-water and oil-water speciation and electrostatic attraction concentration products at different pH conditions. (From Brady, P. V., & Krumhansl, J. L. (2012). A surface complexation model of oil–brine–sandstone interfaces at 100°C: Low salinity waterflooding. *Journal of Petroleum Science and Engineering, 81*, 171–176. https://doi.org/10.1016/j.petrol.2011.12.020.)

between calcite surface groups and oil surface groups is modeled to represent the indirect electrostatic adhesion. In contrast to the sandstone, the carbonate surface is made up of hydrated calcium and carbonate sites. Therefore, there are other surface complexations of the calcite (Table 3.2).

Qiao, Johns, and Li (2016) developed an in-house simulator, PennSim, to model the mechanistic LSWF process incorporating a detailed surface and aqueous multicomponent reaction network involving a variety of adsorption/desorption. The mechanistic model captures the competitive interactions in the system of oil/brine/chalk surface. The study modeled the comprehensive reactions occurred on the interface, as well as the general aqueous reaction and mineral dissolution/precipitation. The mineral reaction is kinetically controlled following the TST rate law. The aqueous reactions and reactions on the surface occur relatively fast and are assumed to be at equilibrium. The reactions occurred on the surface include the various types of surface complexation reactions similar to the study of Brady and Thyne (2016). The surface complexation reactions describe that the calcite surface sites can adsorb the aqueous ions and carboxylate/nitrogen base groups, and then, surface species form. Either EDL or NEDL models are incorporated for the surface complexation modeling. When the change of surface potential is not significant, the NEDL model is used and results in the constant surface potential. Otherwise, the EDL model via Gouy-Chapman theory applies to calculate the surface potential when the surface potential highly changes owing to surface adsorption. For the surface complexation reaction corresponding to Eq. (3.73) on the calcite/water interface, the equilibrium constant is defined as in Eq. (3.74).

$$\equiv \mathrm{CaOH_2}^+ + \mathrm{SO_4^{2-}} \leftrightarrow \mathrm{CaSO_4^-} + \mathrm{H_2O} \tag{3.73}$$

$$K_{eq} = \frac{\exp\left(-\frac{2F\psi_o}{RT}\right)\left[\equiv \mathrm{CaSO_4^-}\right]}{\left[\equiv \mathrm{CaOH_2}^+\right] a_{\mathrm{SO_4^{2-}}}} \tag{3.74}$$

In the interface between oil and water, oil surface complexation reactions are introduced and they are also described by either NEDL or EDL models. This study considers only the following reactions of oil surface complexations of carboxylate base group, not nitrogen gas group, as shown in Eqs. (3.75)–(3.77).

$$\text{-COOH} \leftrightarrow \text{-COO}^- + \mathrm{H}^+ \tag{3.75}$$

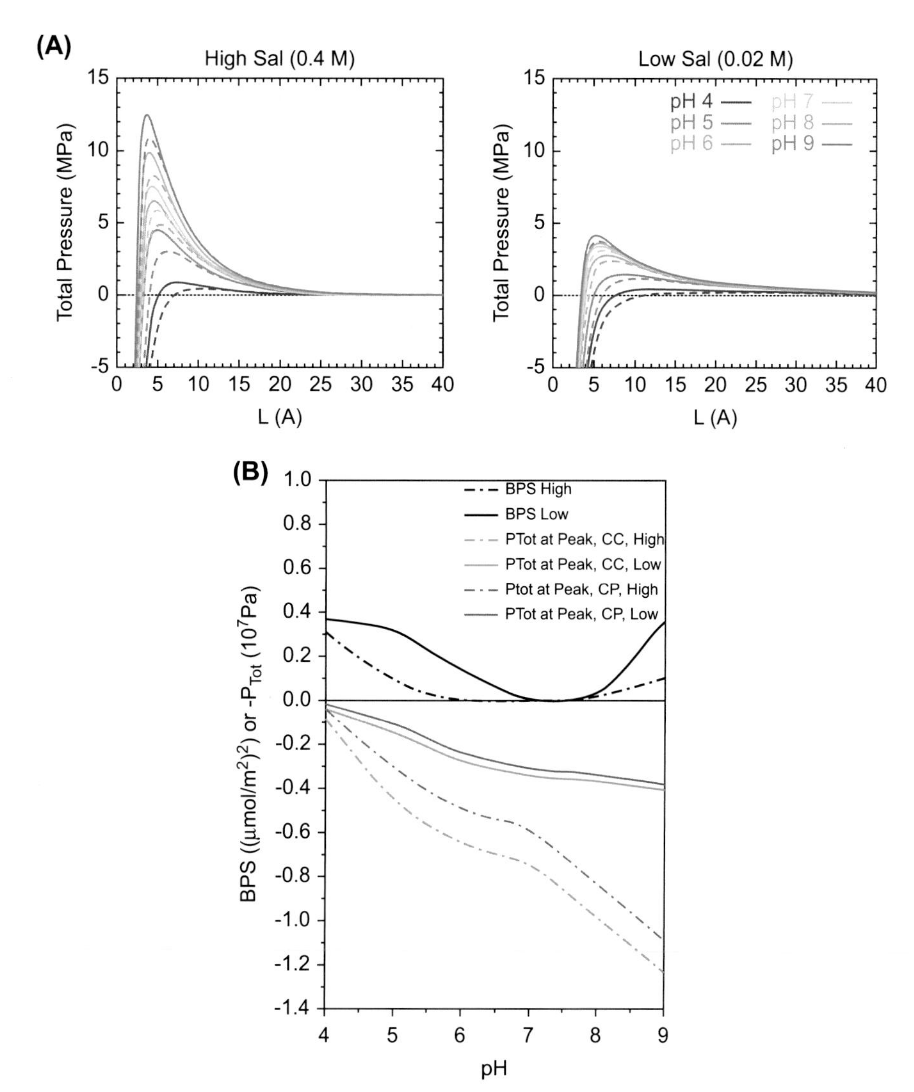

FIG. 3.9 Calculation of **(A)** disjoining pressure and **(B)** bond product sum for low and high salinity conditions. (From Brady, P. V., Morrow, N. R., Fogden, A., Deniz, V., Loahardjo, N., & Winoto. (2015). Electrostatics and the low salinity effect in sandstone reservoirs. *Energy and Fuels, 29*(2), 666–677. https://doi.org/10.1021/ef502474a.)

$$\text{-COOMg}^{+} \leftrightarrow \text{-COO}^{-} + \text{Mg}^{2+} \tag{3.76}$$

$$\text{-COOCa}^{+} \leftrightarrow \text{-COO}^{-} + \text{Ca}^{2+} \tag{3.77}$$

where the prefix "-" represents the species at the oil surface.

Lastly, the modeling of the interface reaction between calcite-water and oil-water is proposed as described in Eq. (3.78). This interface reaction depicts the adsorbed carboxylic group on the calcite surface. The carboxylic group adsorbed on the calcite surface is assumed to be responsible for the wettability modification of carbonate rock. The desorption of the carboxylic group from the calcite surface changes the wettability of calcite surface toward water-wetness. Because the equilibrium constant for this reaction is not well described in the literature, this study used the equilibrium constant as a tuning parameter in the simulation work.

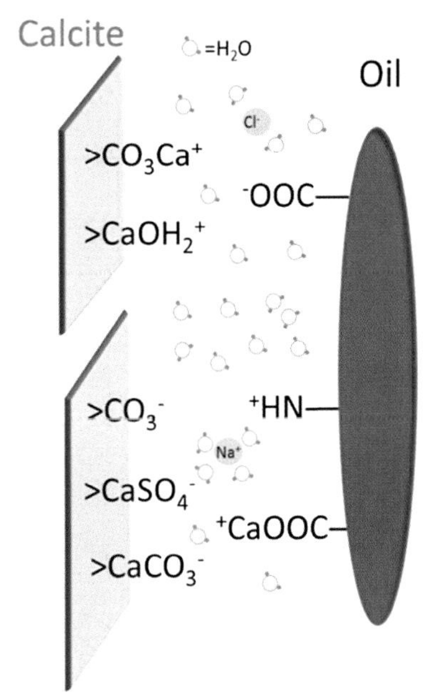

FIG. 3.10 Species on the kaolinite-water surface and oil-water surface (left) and concentration products of electrostatic attraction (right) using surface complexation model. (From Brady, P.V., & Thyne G. (2016). Functional wettability in carbonate reservoirs. *Energy and Fuels*, *30*(11), 9217–9225, https://doi.org/10.1021/acs.energyfuels.6b01895.)

TABLE 3.2
The Speciation of Surface Complexation of Calcite

Reactions
$>CaOH + HCO_3^- \rightleftarrows >CaCO_3^- + H_2O$
$>CaOH_2^+ + SO_4^{2-} \rightleftarrows >CaSO_4^- + H_2O$
$>CO_3H \rightleftarrows >CO_3^- + H^+$
$>CO_3H + Ca^{2+} \rightleftarrows >CO_3Ca^+ + H^+$
$>CO_3H + Mg^{2+} \rightleftarrows >CO_3Mg^+ + H^+$

$$\equiv CaOH_2^+(\text{-COO}^-) \leftrightarrow \equiv CaOH_2^+ + \text{-COO}^- \quad (3.78)$$

where $\equiv CaOH_2^+(\text{-COO}^-)$ indicates the adsorbed carboxylic group on the calcite surface.

The concentration of the adsorbed carboxylic group on the calcite surface is used for the interpolation factor modifying both relative permeability and residual oil saturation. The simulation of wettability modification incorporates the linear modification models for the relative permeability and residual oil saturation. The study simulated the various coreflooding experiments (Austad, Shariatpanahi, Strand, Black, & Webb, 2012; Fathi, Austad, & Strand, 2010; Strand et al., 2008; Yousef et al., 2011) to validate the LSWF model in chalk and limestone cores. Although the mechanistic simulations using NEDL model accurately predict the three experimental measurements (Austad et al., 2012; Fathi et al., 2010; Strand et al., 2008), they fail to model the oil-wet coreflooding of Yousef et al. (2011). Instead, the mechanistic simulation using EDL model successfully reproduces the oil production from the coreflooding experiment. The numerical modeling of wettability modification accurately predicts the experimental measurements. In addition, hypothetical simulations using the history-matched coreflooding (Austad et al., 2012; Strand et al., 2008; Yousef et al., 2011) investigate the role of anhydrite dissolution on the wettability modification. The concentration of adsorbed carboxylic group on the calcite surface is monitored when diluted seawater or diluted formation water is injected into the limestone and chalk (Fig. 3.11). The limestone contains

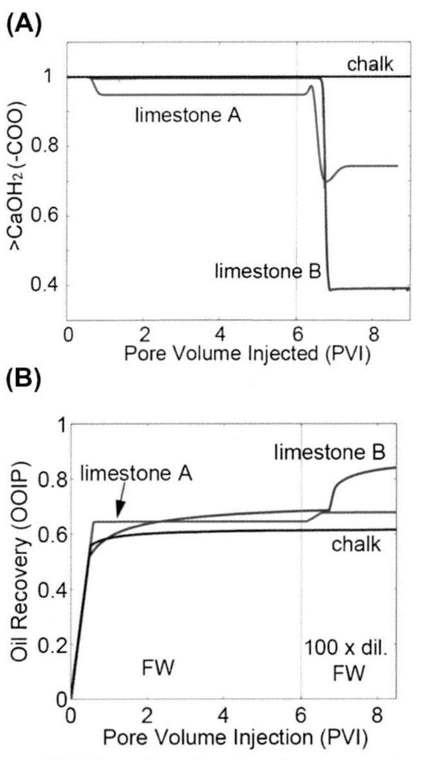

FIG. 3.11 **(A)** Adsorption of carboxylic group on the calcite surface and **(B)** oil recovery when formation water and diluted formation water are injected in chalk and limestone. (From Qiao, C., Johns, R., & Li, L. (2016). Modeling low-salinity waterflooding in chalk and limestone reservoirs. *Energy and Fuels*, *30*(2), 884–895. https://doi.org/10.1021/acs.energyfuels.5b02456.)

the anhydrite mineral, but the chalk does not. The simulations of LSWF result in the increasing or constant adsorption of carboxylic group on the surface for chalk, but decreasing adsorption on the surface for limestone (Fig. 3.11A). Because the desorption of carboxylic group from the surface improves the wetness toward water-wet, the limestone, not chalk, shows the increase of oil recovery (Fig. 3.11B). It is explained that the anhydrite dissolution generates the sulfate ion in water and the increasing sulfate ion leads to the desorption of carboxylic group on the surface due to the reactions of Eqs. (3.73) and (3.78). In addition, the numerical study hypothetically investigates the effects of calcite dissolution on the concentration of adsorbed carboxylic material for the chalk (Fig. 3.12). Despite the equivalent LSWF model, neglecting the calcite dissolution reaction decreases the adsorption of carboxylic material. However, it quantitatively captures the trend of adsorption of carboxylic material as ionic composition of brine changes. Although this study proposed the multiple interactions between crude oil, brine, and solid surface, it neglected the ion bridging interaction owing to lack of experimental and thermodynamic data.

Sanaei, Tavassoli, and Sepehrnoori (2018) from the University of Texas at Austin proposed the application of disjoining pressure based on extended DLVO theory as well as comprehensive geochemical reactions, including the surface complexation reactions to model the LSWF process for both sandstone and carbonate reservoirs (Fig. 3.13). The study upgraded the UTCOMP-IPhreeqc software after the study by Kazemi Nia Korrani et al. (2016). Firstly, the study used the

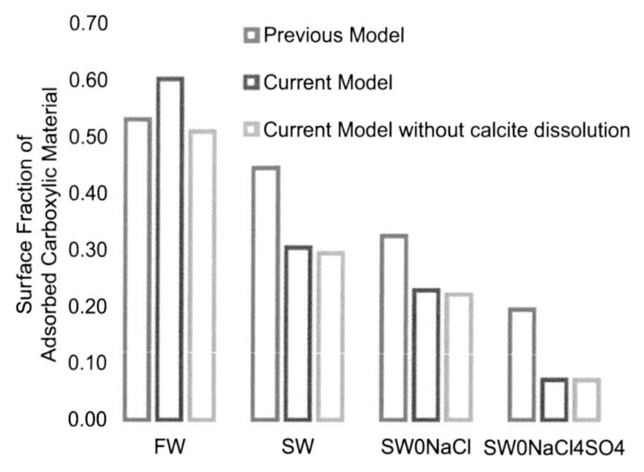

FIG. 3.12 Comparison of the calculated surface fraction of adsorbed carboxylic material for chalk from various models with and without calcite dissolution. (From Qiao, C., Johns, R., & Li, L. (2016). Modeling low-salinity waterflooding in chalk and limestone reservoirs. *Energy and Fuels, 30*(2), 884–895. https://doi.org/10.1021/acs.energyfuels.5b02456.)

PHREEQC software to calculate the two surface complexation models on the oil-water and water-rock interfaces. The surface complexation models determine the ζ-potentials at both oil and rock surfaces, which will be used in the calculation of EDL term in the disjoining pressure based on the extended DLVO theory. The EDL model via Gouy-Chapmann calculates the EDL potential at any point in the system. Secondly, the extended DLVO theory is applied to calculate the interaction potentials and forces across a water film between oil and rock surfaces. With the extended DLVO, the total disjoining pressure is the combination of van der Waal's, EDL, and the structural forces as shown in Eq. (3.79). The negative disjoining pressure indicates the attractive force between the oil and rock surfaces, and the positive disjoining pressure represents the repulsive force between the surfaces.

$$\prod_t(h) = \prod_{\mathrm{VDW}}(h) + \prod_{\mathrm{EDL}}(h) + \prod_s(h) \tag{3.79}$$

where h is the distance between the oil and rock surfaces, i.e., a water film thickness, $\prod_t(h)$ indicates the total disjoining pressure, $\prod_{\mathrm{VDW}}(h)$ is the term of van der Waal's force in the disjoining pressure, $\prod_{\mathrm{EDL}}(h)$ is the term of EDL force in the disjoining pressure, and $\prod_s(h)$ is the term of structural force in the disjoining pressure.

Incorporating the augmented Young-Laplace equation, the disjoining pressure is used to model the wettability modification of LSWF process. At the equilibrium condition, the augmented Young-Laplace equation describes the relationship between disjoining pressure and capillary pressure, i.e., Laplace pressure, as shown in Eq. (3.80). For an infinite thickness of water film, the disjoining pressure is zero and the augmented Young-Laplace equation appears to the conventional Young-Laplace equation. Using the augmented Young-Laplace equation, the contact angle can be derived from the disjoining pressure following Eq. (3.81). The calculated contact angle is used for the wettability modification of LSWF process. The linear modification of relative permeability and capillary pressure uses the interpolation factor as a function of the contact angle.

$$p_c = \prod_t(h) + 2C_m\sigma \tag{3.80}$$

$$1 - \cos\theta = \frac{1}{\sigma}\int_0^{\prod(h_0)} h d\prod = \frac{1}{\sigma}\left(p_c h_0 + \int_{h_0}^{h_\infty} \prod dh\right) \tag{3.81}$$

where C_m is the mean curvature, h_0 is the minimum thickness of water film, and h_∞ is the infinite thickness of water film.

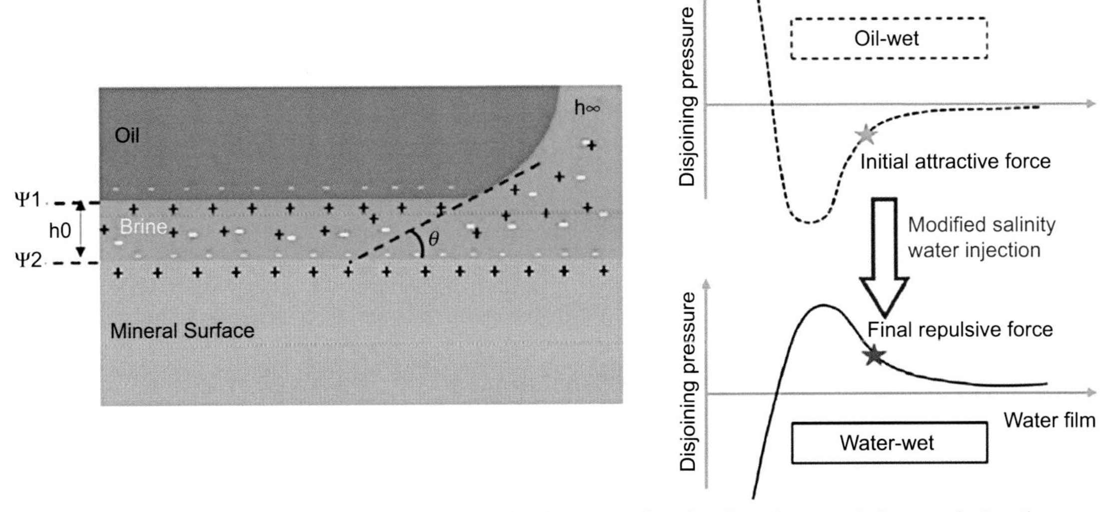

FIG. 3.13 The schematic description of interaction between oil and rock surfaces and change of attractive force to repulsive force owing to variation in electrostatic forces as a result of modified salinity water injection. (From Sanaei, A., & Sepehrnoori, K. (2018). Implication of oil/brine/rock surface interactions in modeling modified salinity waterflooding in carbonate and sandstone reservoirs. In: *Paper presented at the SPE annual technical conference and exhibition, Dallas, Texas, USA, 24–26 September.* https://10.2118/191639-MS.)

The validations of the numerical models are performed by comparing with other experimental works. Firstly, the prediction of contact angle is investigated. The experiments measure contact angle varying pH and composition of brine for both quartz and calcite surfaces. The numerical model using the DLVO theory and augmented Young-Laplace equation does not exactly match the measured contact angle values, but it captures the equal tendency in contact angle change against the experimental results. Secondly, the validation of the surface complexation models is accomplished by investigating and comparing the ζ-potentials at calcite and oil surfaces, respectively, between the simulations and experimental measurements. Tuning the densities of hydrated calcium and carbonate sites of calcite surface, where surface complexation reactions occur, the numerical model predicts the accurate ζ-potential on the calcite surface against the experiments. In addition, the numerical simulation of oil surface complexation model matches the ζ-potential measured on oil surface by slightly tuning the equilibrium constant of the surface complexation reactions. Lastly, the numerical simulation of LSWF coreflooding models the experimental work of Chandrasekhar, Sharma, and Mohanty (2016). The oil recovery and effluent concentrations of ions are compared between simulations and experiments. Except for the calcium concentration, productions from the numerical models are relatively comparable with the experimental measurements. Because the numerical model ignores the kinetics of calcite dissolution, simulation slightly underpredicts the effluent concentration of calcium against the experiments.

Up to date, extensive numerical models have been proposed to describe the LSWF and smart waterflood in sandstone and carbonate reservoirs. There are a few studies to model LSWF following the mechanisms of mineral dissolution or fines migration. Hiorth, Cathles, and Madland (2010) numerically assessed the two potential mechanisms changing carbonate rock wettability: (1) water chemistry changes the charge on the rock surface and (2) changes in the water chemistry could dissolve rock minerals. The numerical simulation investigates how water chemistry influences the surface charge and rock dissolution in pure calcium carbonate rock similar to the Stevns Klint chalk. The first potential mechanism is validated through the numerical simulation of LSWF and ζ-potential calculation. The numerical model uses the bulk aqueous and surface chemistry models as well as mineral reactions of precipitation and dissolution. In the numerical model, the concentration of speciation of surface complexes is predicted. The adsorption of sulfate ion is also estimated and compared with the experimental observation of Strand, Høgnesen, and Austad (2006). Then, the ζ-potential using the Grahame equation is calculated and compared with the experimental measurements of ζ-potential (Strand et al., 2006; Thompson & Pownall, 1989;

Zhang & Austad, 2006). Because oil has the negative potential, the negative surface potential of rock surface indicates the water-wetness in carbonate reservoir. The comparison between surface potentials of oil and calcite can be used to determine rock wettability. Before validating the first mechanism using the ζ-potential calculation, the reliability of ζ-potential calculation is also validated by investigating water stability between carbonate and oil. The stability of water films between carbonate surface and oil indicates the wettability of rock. The stability of water films can be quantified by calculating disjoining pressure. The disjoining pressure discusses the ability of oil to collapse the water film and adhere to the rock surface. In this numerical study, the disjoining pressure is assumed to be only consisted of van der Waals and double layer forces. Using a number of brines, the equivalent wettability of calcite can be interpreted by both ζ-potential and disjoining pressure calculations. Then, the ζ–potential calculation is used to investigate whether the change in ζ-potential or surface potential, i.e., first mechanism, can explain the wettability modification. Using the brines, used in the experiments (Zhang & Austad, 2006; Zhang, Tweheyo, & Austad, 2006, 2007), the ζ-potentials are calculated at increasing temperature. The experimental results of LSWF in carbonate reservoirs show the strong dependency of EOR potential on the temperature. If the surface potential change is the main cause of wettability modification, the surface potential should be strongly sensitive to the temperature. However, the developed model of ζ-potential shows less sensitive to the temperature. Therefore, Hiorth et al. (2010) proposed the calcite mineral dissolution as the main mechanism by calculating calcite dissolution.

The studies (Bedrikovetsky, Siqueira, Furtado, & Souza, 2011; Zeinijahromi, Nguyen, & Bedrikovetsky, 2013) have developed the mechanistic model of LSWF using the fines migration. The numerical model of two-phase flow incorporates the detachment of fine particles, fine migration, and their straining. In the model, the maximum concentration of retained particles as a function of saturation and erosion number, i.e., the ratio between the detaching and attaching torques at the absence of attached fine particles, of porous media, reduces the water permeability. With the assumption of large-scaled approximation, the two-phase flow of fines migration can be similar to the simulation of polymer flood without adsorption. Therefore, the LSWF process involved with fines migration is simulated by using the commercial simulation approach of polymer flood.

However, majority of these studies have developed the numerical models of LSWF assuming the wettability modification as the underlying mechanism. They interpret the mechanism of wettability modification in terms of the geochemical reactions, electrokinetics, etc. Therefore, the models of LSWF process incorporate the calculations of comprehensive geochemical reactions in the crude oil/brine/rock system and/or electrokinetics in the system. Although the extensive studies provide the simulation results of LSWF comparable with the experimental results, only a few studies have been published for the LSWF modeling of the field-scaled trials.

FIELD-SCALED MODELING

Kazemi Nia Korrani et al. (2016) numerically modeled the interwell field trials of LSWF in Endicott field on the North Slope, which have been reported by Seccombe, Lager, Webb, Jerauld, and Fueg (2008) and Seccombe et al. (2010). In the field test, LSWF is applied into Well 3–35 after injection of produced water. The production and response from Well 3–37 are monitored. From the interwell test, successfully, the reduction of residual oil saturation with 0.14 units is observed. Kazemi Nia Korrani et al. (2016) constructed a multilayer model of the target reservoir using the UTCOMP-IPhreeqc simulator. The effluent ions, water-cut, pH, and alkalinity of numerical simulations are compared with the historical data of field tests. Because the simulator of UTCOMP-IPhreeqc can model the reactions of the aqueous/rock geochemistry because of soluble hydrocarbon components and surface complexation with exchanger, the numerical simulation of LSWF accurately reproduces the historical data of field tests. The numerical simulations cover the dissolution of soluble hydrocarbon components as well as CO_2 into water, which influences the pH of brine because of the aqueous and mineral reactions. Modeling the reactions leads to the correct trend for pH of the produced water. In addition, the hypothetical simulations of LSWF process with and without the surface complexation reaction are carried out. The results of both simulations are compared with the measured data of the field tests. The LSWF model without the reaction shows a significant discrepancy against the measured data of the field tests, especially in terms of alkalinity and iron concentrations (Fig. 3.14). Measured alkalinity and iron concentrations are higher than the results from the simulation neglecting the reactions. However, the LSWF model with the reactions results in the alkalinity and iron concentrations comparable with the measured data. In the study, it is also hypothesized that the underlying mechanism of LSWF process is the wettability modification, which is a function of total ionic strength. The numerical model

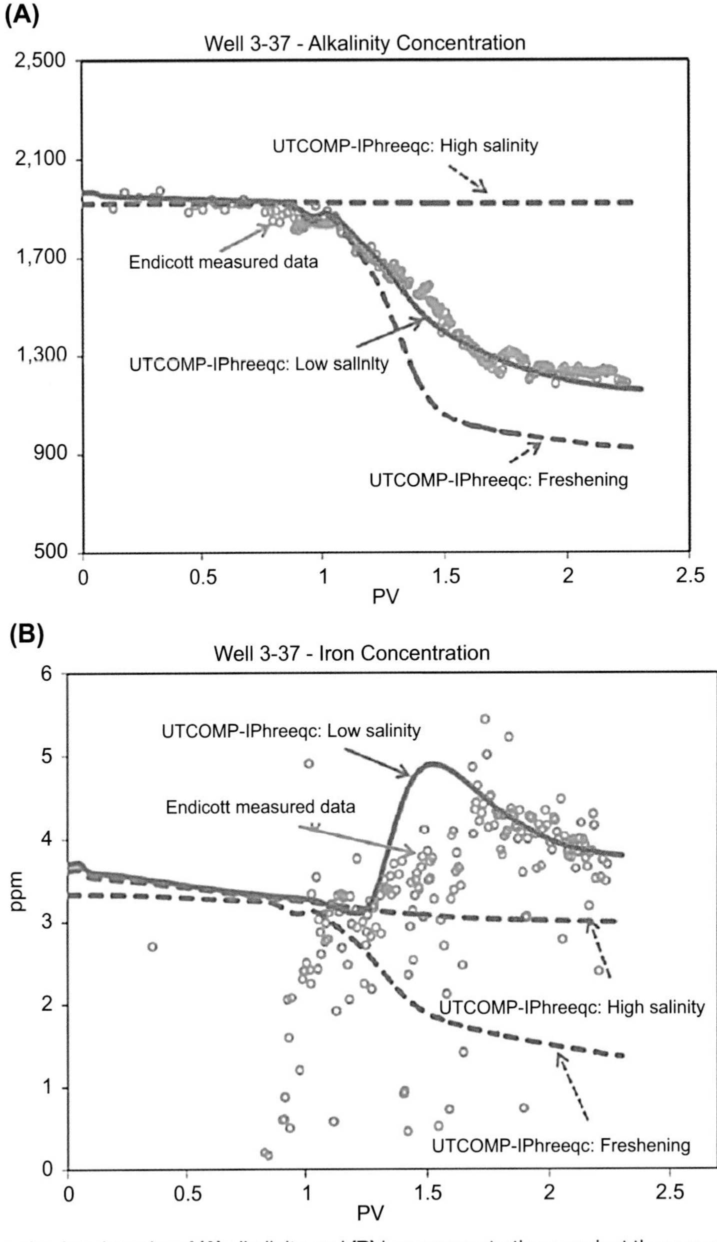

FIG. 3.14 The simulated results of **(A)** alkalinity and **(B)** iron concentrations against the measured data of the interwell field trial. (From Kazemi Nia Korrani, A., Jerauld, G. R., & Sepehrnoori, K. (2016). Mechanistic modeling of low-salinity waterflooding through coupling a geochemical package with a compositional reservoir simulator. *SPE Reservoir Evaluation and Engineering, 19*(1), 142–162. https://doi.org/10.2118/169115-PA.)

of LSWF accurately matches the water-cut of the field trials in Endicott field.

Yousef, Al-Saleh, et al. (2012) and Yousef, Liu, et al. (2012) also reported the field trials of SWCTT at Well A and Well B to analyze the performance of LSWF process in a carbonate reservoir and numerical simulations of the SWCTT and LSWF processes. The numerical simulations use the two important mechanisms of SWCTT test and LSWF process in carbonate reservoirs. The SWCTT test involves the injection of tracers (ethyl acetate, methyl alcohol, and isopropyl alcohol) and a couple of reactions including partitioning reaction and hydrolysis. There is a partitioning reaction, at which the primary tracer of ester, i.e., ethyl acetate (EtAC), dissolves in both oil and water. Some ester dissolved in water hydrolyzes to form a secondary or product tracer, i.e., ethyl alcohol (EtOH). The multicomponent and multiphase flow simulation of SWCTT process has to use the reactions. In addition, the mechanism of LSWF process, wettability modification, has to be implemented in the numerical simulation. The mechanistic approach of LSWF modeling is associated with the comprehensive geochemical reactions as well as wettability modification modeling. The numerical simulations of LSWF process use the input parameters determined by the previous experiments (Yousef, Al-Saleh, et al., 2012; Yousef et al., 2011). It is assumed that the LSWF improves the wettability of carbonate rocks modifying relative permeability and capillary pressure.

The field trial design for Well A is consisted of three stages. In the first stage, sufficient field seawater is injected to establish the residual oil saturation near Well A. Afterward, the SWCTT test is deployed to estimate the residual oil saturation after seawater injection. In the second stage, seawater injection and succeeding SWCTT test are repeated to ensure the residual oil saturation. The last stage is incorporated with the LSWF using diluted seawater by a factor of 10, and then, the SWCTT test is performed to measure the residual oil saturation. Numerical simulations are carried out for the field trial design of three stages. From the results of the numerical simulation, the reduction in residual oil saturation is estimated by 6%–7% in the vicinity of Well A. The prediction of simulations is similar to the measurement of the real field trial implementation for Well A. For the Well B, the field trial is differently designed to demonstrate the impact of different versions of LSWF on the residual oil saturation reduction. The field trial design for Well B also has three stages. In the first stage, the seawater is flooded through Well B and SWCTT is performed to measure the residual oil saturation. In the second stage, the reduction of residual oil saturation by LSWF, which injects the diluted seawater by a factor of 2, is investigated. The SWCTT process measures the residual oil saturation after the LSWF. In the third stage, another LSWF process is deployed and then SWCTT is performed. The second LSWF uses the diluted seawater by a factor of 10. From the results of numerical simulations, it is shown that LSWF with two-times-diluted seawater reduces residual oil saturation by 3%. LSWF with 10-times-diluted seawater decreases the residual oil saturation by 4%–6% with an uncertainty. There is a slight discrepancy between the numerical simulation and the field measurement because of uncertainty of implementations of SWCTT for Well B.

Up to date, only a few studies have reported the field-scaled simulations of LSWF process in sandstone and carbonate reservoirs. The field-scaled simulations are developed from the core-scaled numerical simulation and experiments. In addition, the field-scaled simulation of Yousef, Al-Saleh, et al. (2012) and Yousef, Liu, et al. (2012) reliably predicted real field implementation and demonstrated the feasibility of LSWF process in the field.

REFERENCES

Al-Shalabi, E. W., & Sepehrnoori, K. (2017). *Low salinity and engineered water injection for sandstones and carbonate reservoirs*. Cambridge, MA: Gulf Professional Publishing, an imprint of Elsevier.

Al-Shalabi, E. W., Luo, H., Delshad, M., & Sepehrnoori, K. (2015a). Single-well chemical tracer modeling of low salinity water injection in carbonates. In *Paper presented at the SPE western regional meeting, Garden Grove, California, USA, 27–30 April*. https://doi.org/10.2118/173994-MS.

Al-Shalabi, E. W., Sepehrnoori, K., Delshad, M., & Pope, G. (2015b). A novel method to model low-salinity-water injection in carbonate oil reservoirs. *SPE Journal, 20*(5), 1154–1166. https://doi.org/10.2118/169674-PA.

Al-Shalabi, E. W., Sepehrnoori, K., & Pope, G. (2015c). Mechanistic modeling of oil recovery due to low salinity water injection in oil reservoirs. In *Paper presented at the SPE middle east oil & gas show and conference, Manama, Bahrain, 8–11 March*. https://doi.org/10.2118/172770-MS.

Al-Shalabi, E. W., Sepehrnoori, K., Pope, G., & Mohanty, K. (2014). A fundamental model for predicting oil recovery due to low salinity water injection in carbonate rocks. In *Paper presented at the SPE energy resources conference, port of Spain, Trinidad and Tobago, 9–11 June*. https://doi.org/10.2118/169911-MS.

Appelo, C. A. J., & Postma, D. (1999). *Geochemistry, groundwater and pollution* (4th corrected print ed.). Rotterdam, Brookfield, VT: Balkema.

Austad, T., Shariatpanahi, S. F., Strand, S., Black, C. J. J., & Webb, K. J. (2012). Conditions for a low-salinity enhanced oil recovery (EOR) effect in carbonate oil reservoirs. *Energy and Fuels, 26*(1), 569–575. https://doi.org/10.1021/ef201435g.

Bakker, R. J. (2003). Package FLUIDS 1. computer programs for analysis of fluid inclusion data and for modelling bulk fluid properties. *Chemical Geology, 194*(1), 323. https://doi.org/10.1016/S0009-2541(02)00268-1.

Bedrikovetsky, P., Siqueira, F. D., Furtado, C. A., & Souza, A. L. S. (2011). Modified particle detachment model for colloidal transport in porous media. *Transport in Porous*

Media, 86(2), 353–383. https://doi.org/10.1007/s11242-010-9626-4.

Bethke, C. (1996). *Geochemical reaction modeling: Concepts and applications*. New York: Oxford University Press.

Brady, P. V., & Krumhansl, J. L. (2012). A surface complexation model of oil–brine–sandstone interfaces at 100°C: Low salinity waterflooding. *Journal of Petroleum Science and Engineering, 81*, 171–176. https://doi.org/10.1016/j.petrol.2011.12.020.

Brady, P. V., Krumhansl, J. L., & Mariner, P. E. (2012). Surface complexation modeling for improved oil recovery. In *Paper presented at the SPE improved oil recovery symposium, Tulsa, Oklahoma, USA, 14–18 April*. https://doi.org/10.2118/153744-MS.

Brady, P. V., Morrow, N. R., Fogden, A., Deniz, V., Loahardjo, N., & Winoto. (2015). Electrostatics and the low salinity effect in sandstone reservoirs. *Energy and Fuels, 29*(2), 666–677. https://doi.org/10.1021/ef502474a.

Brady, P. V., & Thyne, G. (2016). Functional wettability in carbonate reservoirs. *Energy and Fuels, 30*(11), 9217–9225. https://doi.org/10.1021/acs.energyfuels.6b01895.

Chandrasekhar, S., Sharma, H., & Mohanty, K. K. (2016). Wettability alteration with brine composition in high temperature carbonate rocks. In *Paper presented at the SPE annual technical conference and exhibition, Dubai, UAE, 26–28 September*. https://doi.org/10.2118/181700-MS.

Dang, C. T. Q., Nghiem, L. X., Chen, Z. J., & Nguyen, Q. P. (2013). Modeling low salinity waterflooding: Ion exchange, geochemistry and wettability alteration. In *Paper presented at the SPE annual technical conference and exhibition, New Orleans, Louisiana, USA, 30 September–2 October*. https://doi.org/10.2118/166447-MS.

Drever, J. I. (1997). *The geochemistry of natural waters: Surface and groundwater environments* (3rd ed.). Upper Saddle River, NJ: Prentice Hall.

Eyring, H. (1935). The activated complex in chemical reactions. *The Journal of Chemical Physics, 3*(2), 107–115. https://doi.org/10.1063/1.1749604.

Fathi, S. J., Austad, T., & Strand, S. (2010). "Smart water" as a wettability modifier in chalk: The effect of salinity and ionic composition. *Energy and Fuels, 24*(4), 2514–2519. https://doi.org/10.1021/ef901304m.

Fjelde, I., Asen, S. M., & Omekeh, A. V. (2012). Low salinity water flooding experiments and interpretation by simulations. In *Paper presented at the SPE improved oil recovery symposium, Tulsa, Oklahoma, USA, 14–18 April*. https://doi.org/10.2118/154142-MS.

Garcia, J. E. (2001). *Density of aqueous solutions of CO_2*. Lowrence Livermore National Laboratory.

Harvey, A. H. (1996). Semiempirical correlation for henry's constants over large temperature ranges. *AIChE Journal, 42*(5), 1491–1494. https://doi.org/10.1002/aic.690420531.

Havre, T. E., Sjöblom, J., & Vindstad, J. E. (2003). Oil/water-partitioning and interfacial behavior of naphthenic acids. *Journal of Dispersion Science and Technology, 24*(6), 789–801. https://doi.org/10.1081/DIS-120025547.

Hiorth, A., Cathles, L. M., & Madland, M. V. (2010). The impact of pore water chemistry on carbonate surface charge and oil wettability. *Transport in Porous Media, 85*(1), 121. https://doi.org/10.1007/s11242-010-9543-6.

Jerauld, G. R., Webb, K. J., Lin, C.-Y., & Seccombe, J. (2006). Modeling low-salinity waterflooding. In *Paper presented at the SPE annual technical conference and exhibition, San Antonio, Texas, USA, 24–27 September*. https://doi.org/10.2118/102239-MS.

Jerauld, G. R., Webb, K. J., Lin, C.-Y., & Seccombe, J. C. (2008). Modeling low-salinity waterflooding. *SPE Reservoir Evaluation and Engineering, 11*(6), 1000–1012. https://doi.org/10.2118/102239-PA.

Kazemi Nia Korrani, A., Jerauld, G. R., & Sepehrnoori, K. (2016). Mechanistic modeling of low-salinity waterflooding through coupling a geochemical package with a compositional reservoir simulator. *SPE Reservoir Evaluation and Engineering, 19*(1), 142–162. https://doi.org/10.2118/169115-PA.

Kazemi Nia Korrani, A., Sepehrnoori, K., & Delshad, M. (2013). A novel mechanistic approach for modeling low salinity water injection. In *Paper presented at the SPE annual technical conference and exhibition, New Orleans, Louisiana, USA, 30 September–2 October*. https://doi.org/10.2118/166523-MS.

Kozaki, C. (2012). *Efficiency of low salinity polymer flooding in sandstone cores* (Master's thesis). The University of Texas at Austin.

Li, Y.-K., & Nghiem, L. X. (1986). Phase equilibria of oil, gas and water/brine mixtures from a cubic equation of state and henry's law. *The Canadian Journal of Chemical Engineering, 64*(3), 486–496. https://doi.org/10.1002/cjce.5450640319.

Mohanty, K. K., & Chandrasekhar, S. (2013). Wettability alteration with brine composition in high temperature carbonate reservoirs. In *Paper presented at the SPE annual technical conference and exhibition, New Orleans, Louisiana, USA, 30 September–2 October*. https://doi.org/10.2118/166280-MS.

Nghiem, L., Sammon, P., Grabenstetter, J., & Ohkuma, H. (2004). Modeling CO_2 storage in aquifers with a fully-coupled geochemical EOS compositional simulator. In *Paper presented at the SPE/DOE symposium on improved oil recovery, Tulsa, Oklahoma, 17–21 April*. https://doi.org/10.2118/89474-MS.

Omekeh, A. V., Friis, H. A., Fjelde, I., & Evje, S. (2012). Modeling of ion-exchange and solubility in low salinity water flooding. In *Paper presented at the SPE improved oil recovery symposium, Tulsa, Oklahoma, USA, 14–18 April*. https://doi.org/10.2118/154144-MS.

Parkhurst, D. L., & Appelo, C. A. J. (1999). *User's guide to PHREEQC (version 2): A computer program for speciation, batch-reaction, one-dimensional transport, and inverse geochemical calculations, water-resources investigations report*. Denver, Colo: U.S. Department of the Interior, U.S. Geological Survey.

Peng, D.-Y., & Robinson, D. B. (1976). A new two-constant equation of state. *Industrial and Engineering Chemistry Fundamentals, 15*(1), 59–64. https://doi.org/10.1021/i160057a011.

Pope, G. A., Wu, W., Narayanaswamy, G., Delshad, M., Sharma, M. M., & Wang, P. (2000). Modeling relative permeability effects in gas-condensate reservoirs with a new trapping model. *SPE Reservoir Evaluation and Engineering, 3*(2), 171–178. https://doi.org/10.2118/62497-PA.

Qiao, C., Johns, R., & Li, L. (2016). Modeling low-salinity waterflooding in chalk and limestone reservoirs. *Energy and Fuels, 30*(2), 884–895. https://doi.org/10.1021/acs.energyfuels.5b02456.

Rivet, S. M. (2009). *Coreflooding oil displacements with low salinity brine* (Master's thesis). University of Texas at Austin.

Sanaei, A., Tavassoli, S., & Sepehrnoori, K. (2018). Investigation of modified water chemistry for improved oil recovery: Application of DLVO theory and surface complexation model. In *Paper presented at the SPE western regional meeting, Garden Grove, California, USA, 22–26 April.* https://doi.org/10.2118/190017-MS.

Saul, A., & Wagner, W. (1987). International equations for the saturation properties of ordinary water substance. *Journal of Physical and Chemical Reference Data, 16*(4), 893–901. https://doi.org/10.1063/1.555787.

Seccombe, J., Lager, A., Jerauld, G., Jhaveri, B., Todd, B., Bassler, S., et al. (2010). Demonstration of low-salinity EOR at interwell scale, Endicott field, Alaska. In *Paper presented at the SPE improved oil recovery symposium, Tulsa, Oklahoma, USA, 24–28 April.* https://doi.org/10.2118/129692-MS.

Seccombe, J. C., Lager, A., Webb, K. J., Jerauld, G., & Fueg, E. (2008). Improving wateflood recovery: LoSalTM EOR field evaluation. In *Paper presented at the SPE symposium on improved oil recovery, Tulsa, Oklahoma, USA, 20–23 April.* https://doi.org/10.2118/113480-MS.

Søreide, I., & Whitson, C. H. (1992). Peng-robinson predictions for hydrocarbons, CO_2, N_2, and H_2S with pure water and NaCI brine. *Fluid Phase Equilibria, 77*, 217–240. https://doi.org/10.1016/0378-3812(92)85105-H.

Strand, S., Austad, T., Puntervold, T., Høgnesen, E. J., Olsen, M., & Barstad, S. M. F. (2008). "Smart water" for oil recovery from fractured limestone: A preliminary study. *Energy and Fuels, 22*(5), 3126–3133. https://doi.org/10.1021/ef800062n.

Strand, S., Høgnesen, E. J., & Austad, T. (2006). Wettability alteration of carbonates—effects of potential determining ions (Ca^{2+} and SO_4^{2-}) and temperature. *Colloids and Surfaces A: Physicochemical and Engineering Aspects, 275*(1), 110. https://doi.org/10.1016/j.colsurfa.2005.10.061.

Thompson, D. W., & Pownall, P. G. (1989). Surface electrical properties of calcite. *Journal of Colloid and Interface Science, 131*(1), 7482. https://doi.org/10.1016/0021-9797(89)90147-1.

Yousef, A. A., Al-Saleh, S., & Al-Jawfi, M. S. (2012). Improved/enhanced oil recovery from carbonate reservoirs by tuning injection water salinity and ionic content. In *Paper presented at the SPE improved oil recovery symposium, Tulsa, Oklahoma, USA, 14–18 April.* https://doi.org/10.2118/154076-MS.

Yousef, A. A., Al-Saleh, S. H., Al-Kaabi, A., & Al-Jawfi, M. S. (2011). Laboratory investigation of the impact of injection-water salinity and ionic content on oil recovery from carbonate reservoirs. *SPE Reservoir Evaluation and Engineering, 14*(5), 578–593. https://doi.org/10.2118/137634-PA.

Yousef, A. A., Liu, J. S., Blanchard, G. W., Al-Saleh, S., Al-Zahrani, T., Al-Zahrani, R. M., et al. (2012). Smart waterflooding: Industry. In *Paper presented at the SPE annual technical conference and exhibition, San Antonio, Texas, USA, 8–10 October.* https://doi.org/10.2118/159526-MS.

Zeinijahromi, A., Nguyen, T. K. P., & Bedrikovetsky, P. (2013). Mathematical model for fines-migration-assisted waterflooding with induced formation damage. *SPE Journal, 18*(3), 518–533. https://doi.org/10.2118/144009-PA.

Zhang, P., & Austad, T. (2006). Wettability and oil recovery from carbonates: Effects of temperature and potential determining ions. *Colloids and Surfaces A: Physicochemical and Engineering Aspects, 279*(1), 179–187. https://doi.org/10.1016/j.colsurfa.2006.01.009.

Zhang, P., Tweheyo, M. T., & Austad, T. (2006). Wettability alteration and improved oil recovery in Chalk: the effect of calcium in the presence of sulfate. *Energy and Fuels, 20*(5), 2056–2062. https://doi.org/10.1021/ef0600816.

Zhang, P., Tweheyo, M. T., & Austad, T. (2007). Wettability alteration and improved oil recovery by spontaneous imbibition of seawater into chalk: Impact of the potential determining ions Ca^{2+}, Mg^{2+}, and SO_4^{2-}. *Colloids and Surfaces A: Physicochemical and Engineering Aspects, 301*(1), 199208. https://doi.org/10.1016/j.colsurfa.2006.12.058.

CHAPTER 4

Hybrid Chemical EOR Using Low-Salinity and Smart Waterflood

ABSTRACT

This chapter describes the application of low-salinity waterflood (LSWF) into various chemical enhanced oil recovery (EOR) processes, including polymer flood, gel treatment, surfactant flood, and alkaline flood. The hybrid process can secure the advantages of both LSWF and chemical EOR improving sweep and displacement efficiencies. Because the properties of chemical additives are sensitive to the salinity of brine, the synergetic effects can be introduced by the hybrid process of LSWF and chemical EOR. Extensive experiments and numerical simulations have explored the applications of the hybrid EOR process and quantified the underlain synergy. This chapter briefly discusses the important backgrounds of chemical EOR and then introduces the research studies of a variety of hybrid chemical EOR using low-salinity and smart waterflood.

Although the exact mechanism of low-salinity and smart waterflood is still under investigation, the applications of low-salinity and smart waterflood into other enhanced oil recovery (EOR) processes have been proposed and evaluated. This section describes the investigations of hybrid processes of the LSWF and chemical EOR processes, including polymer flood/gel treatment, surfactant flood, and alkali flood. It addresses the synergy of the hybrid process to enhance oil recovery. The synergy might introduce effects of LSWF, which are mainly wettability modification and/or stability improvement of chemical additives.

POLYMER FLOOD/GEL TREATMENT

Backgrounds of Polymer Flood/Gel Treatment

Before the description of LSWF application into polymer flood and gel treatment, it is necessary to discuss the fundamental theories of polymer flood and gel treatment, which are related to the synergetic effects of hybrid process. This discussion is referred from the previous works (Sheng, 2011; Sorbie, 1991). The two types of polymers, synthetic polymer and biopolymer, are commonly used in the polymer and gel EOR applications. The hydrolyzed polyacrylamide (HPAM) is the widely used synthetic polymer and xanthan is one of the biopolymers. The addition of these polymers contributes to viscosifying the displacing fluid, which is of interest to the EOR application of polymer. The degree of viscosifying the fluid is closely associated with the physical properties of the polymer, including flow behavior, adsorption/retention, and mechanical, chemical, and thermal stabilities as well as polymer concentration. The chemical structure of polymer determines the physical properties. Because each polymer has different chemical structure, i.e., flexible coil structure of HPAM and rigid molecule structure of xanthan, its physical properties vary.

Polymer viscosity

The polymers are added to the injecting brine to increase the viscosity of the driving fluid, which in turn improves the mobility ratio between displacing and displaced fluids. When water or polymeric solution displaces oil, the mobility ratio and mobility are defined as in Eqs. (4.1) and (4.2). Generally, the mobility ratio equal to or less than one indicates the favorable condition for displacement and higher than one means the unfavorable condition.

$$M = \frac{\lambda_D}{\lambda_d} \tag{4.1}$$

$$\lambda_j = k\left(\frac{k_{rj}}{\mu_j}\right) \tag{4.2}$$

where M is the mobility ratio of the displacing phase, λ_D, to the displaced phase, λ_d; λ_j is the mobility of phase j; and the μ_j is the viscosity of phase j.

The viscosity of a polymeric solution is determined by the molecular size, concentration of polymer, and extension of polymer in the solution. Conventionally, the lager molecular size of polymer has higher viscosity. There are a number of quantities to describe the

Hybrid Enhanced Oil Recovery using Smart Waterflooding. https://doi.org/10.1016/B978-0-12-816776-2.00004-0

viscosity: relative viscosity, specific viscosity, reduced viscosity, inherent viscosity, and intrinsic viscosity. The quantity of intrinsic viscosity is the fundamental indication of molecular weight and it is independent of polymer concentration. Eq. (4.3) defines the intrinsic viscosity as the specific viscosity at the infinite dilution concentration. Many equations have been proposed to quantify the relationship between the intrinsic viscosity and molecular weight of polymer. The equations explain that the viscosity is proportional to the molecular weight of polymer, because the polymeric solutions used in polymer EOR have relatively low concentration, in which interactions between polymer molecules are negligible. The relationships proposed by the studies (Chauveteau, 1982; Huggins, 1942; Kraemer, 1938) are commonly used to describe in the light of the intrinsic viscosity of polymer. Huggins (1942) developed the relationship between the specific viscosity and polymer concentration at the low concentration as in Eq. (4.4). Kraemer (1938) formulated the relation between the inherent viscosity and intrinsic viscosity as in Eq. (4.5).

$$[\eta] = \lim_{c_{\text{poly}} \to \infty} \left(\frac{\eta_{sp}}{c_{\text{poly}}} \right) \tag{4.3}$$

$$\frac{\eta_{sp}}{c_{\text{poly}}} = [\eta] + k'[\eta]^2 c_{\text{poly}} \tag{4.4}$$

$$\eta_I = \frac{\ln \eta_r}{c_{\text{poly}}} = [\eta] - k''[\eta]^2 c_{\text{poly}} \tag{4.5}$$

where c_{poly} indicates the concentration of polymer; η is the non-Newtonian fluid's viscosity: if the viscosity η is independent of the shear rate, then it is equal to the Newtonian fluid's viscosity μ; $[\eta]$ is the intrinsic viscosity; η_{sp} is the specific viscosity; k' is the Huggins constant; η_I is the inherent viscosity; η_r is the relative viscosity; and k'' is the constant.

The intrinsic viscosity is also sensitive to the molecular expansion of the polymers as well as the molecular weight. Most of the real polymers have nonlinear chains of polymer. The real end-to-end size of polymer has a discrepancy to the apparent size of polymer in a solution. This difference differentiates the intrinsic viscosity of real polymer from that of linear structure of polymer. The size of polymer chain is also affected by the solvent. Both HPAM and xanthan are not neutral. They are polyelectrolytes with a number of negative charges, which introduce the electrostatic repulsion between the different parts of the molecule. Therefore, the molecular expansion of polymer is influenced by the solutions.

Salinity, hardness, pH, temperature, oxidation reduction, and ironic ions

Both HPAM and xanthan undergo the degradations by salinity, hardness, pH, oxidation reduction, and ironic ions, which are related to the chemical stability. They are flexible polyelectrolyte. The salt can screen the charged molecules of polymer, and the charged molecules of polymer in a solution would be surrounded by the opposite charged ions. Therefore, the polyelectrolyte polymer molecules would expand at the low salt concentration because of more mutual repulsion of the charges along its chain and contract at the high salt concentration owing to less repulsion. The expansion in the molecular size of polymer increases the viscosity of polymeric solution. For the HPAM, the divalent cations, i.e., hardness, have more significant screening effect on the negatively charged molecules of polymer compared with the monovalent cation. As HPAM undergoes more hydrolysis, the negatively charged carboxyl group of HPAM interacts with the divalent cations. It results in the decreasing viscosity of polymeric solution, formation of gels, or precipitation. The viscosity of xanthan is less sensitive to the salinity and hardness compared with HPAM. The pH also influences the viscosity of HPAM. The polymer is neutral at low pH condition but completely charged at high pH condition. Lowering pH neutralizes the polyelectrolyte HPAM polymer and leads to the less expansion of polymer chain because of less electrostatic effects. The effect of pH becomes more dominant in low-salinity solutions. In terms of temperature, the high temperature destabilizes both HPAM and xanthan polymers and reduces the viscosities, i.e., thermal degradation of polymer. In addition, the oxidative degradation of polyacrylamide polymer varies with the temperature. At the low temperature, the effect of dissolved oxygen on viscosity of HPAM is not significant. However, even small amount of dissolved oxygen provides the substantial reduction of the viscosity at the high temperature condition. Lastly, the high enough concentration of ferric ion (Fe^{3+}) cross-links the HPAM to form insoluble gel. The loss of viscosity is significant because of insoluble gel formation.

Shear rate

In contrast to the Newtonian fluid of water, polymeric solution suffers a non-Newtonian behavior by the shear stress. It is understood as the mechanical stability of polymer. Although the Newtonian fluid of water, theoretically, has a constant viscosity regardless of shear stress, the non-Newtonian fluid of polymeric solution has shear rate-dependent viscosity, i.e., the viscosity changes corresponding to the shear rate. As the polymer EOR deployment uses the polymeric solution at low concentration of polymer, the pseudoplastic or shear-thinning behaviors are applicable to the vast majority of polymeric solution in diluted homogeneous regimes. Generally, the shear-thinning polymer shows the decreasing viscosity with an increase in shear stress or shear rate. The analytical form of the power law, described in Eq. (4.6), commonly describes the shear-thinning behavior between the viscosity of polymeric solution and shear rate (Bird, 1960). The two-parameter equation of the power law is also known as the Ostwald-de Waele model. The model is applicable to the pseudoplastic regime, not to the high and low shear rate regimes.

$$\eta(\dot{\gamma}) = K\dot{\gamma}^{n-1} \tag{4.6}$$

where $\dot{\gamma}$ is the shear rate; K is the flow consistency index; and n is the flow behavior index.

The Carreau equation of Eq. (4.7) describes the shear-thinning behavior more accurately in the whole regime (Bird, Armstrong & Hassage, 1987; Carreau, 1972).

$$\eta(\dot{\gamma}) = \eta_\infty + (\eta_0 - \eta_\infty)[1 + (A\dot{\gamma})^\alpha]^{\frac{n-1}{\alpha}} \tag{4.7}$$

where η_0 is the viscosity at very low shear rate; η_∞ is the viscosity of limiting value at high shear rate; A and n are the polymer-specific empirical constants; and α is generally equal to 2.

There are other models to describe the pseudoplastic rheology of polymeric solution. Often, the polymeric solutions show an increasing apparent viscosity in high shear stress condition. This relation of an increasing apparent viscosity with an increase in shear stress or shear rate is shear-thickening or dilatant behavior. A couple of mechanisms have been proposed (Clarke, Howe, Mitchell, Staniland, & Hawkes, 2016; Delshad et al., 2008; Seright, Fan, Wavrik, de Carvalho Balaban, 2011), and they explain that viscoelastic of polymer improves the displacement efficiency reducing residual oil. The research team from the University of Texas at Austin developed the mechanistic model, which incorporates the Carreau equation for shear-thinning behavior, to represent viscoelastic behavior. The proposed model is the summation of shear-thinning and thickening behaviors of polymeric solution (Delshad et al., 2008). Incorporating the viscoelastic model and experiments, the rheology database of the synthetic polymers has been constructed. Lee, Kim, Huh, and Pope (2009) measured the viscosity of three synthetic polymers of Flopaam 3330s, Flopaam 3630s, and AN-125, in the laboratory experiments. The study matched the Carreau model, which is the shear-thinning part of the mechanistic model, as a function of polymer concentration, salinity, hardness, and temperature to the measured data in the shear-thinning regime (Fig. 4.1). Kim, Lee, Ahn, Huh, and Pope (2010) developed the shear-thickening part of the mechanistic model to correspond the rheology of synthetic polymers as a function of the same parameters (Fig. 4.2).

Retention

There are three types of polymer retention: adsorption, mechanical trapping, and hydrodynamic retention. Although the adsorption occurs in both stagnant and transport regimes, the mechanical trapping and hydrodynamic retention only occur in the transport regime within the porous media. The mechanical trapping describes the larger molecules to be stuck in the narrow channels or pores. Therefore, it highly depends on the pore size distribution. The hydrodynamic retention is easily thought to be trapped temporarily in the stagnant flow region, i.e., close to the porous media, because of the hydrodynamic drag force. Because of the hydrodynamic retention, higher concentration of polymer can exist in this region compared with that in the bulk of the injecting solutions. The mechanical trapping can be avoided through screening test. The hydrodynamic retention is relatively minor and can be ignored in the practical applications. Because the adsorption of polymer is significant compared with the mechanical trapping and hydrodynamic retention, adsorption is the major interest in the polymer retention. The adsorption occurs by the interaction between polymer molecules and solid surface. The main contribution on the interaction is the physical adsorption by van der Waal's and hydrogen bonding rather than chemisorption. The solid surface with a larger surface area leads to the more adsorption of polymer and the significant removal of polymer from the bulk solution. Although the hydrodynamic retention is conventionally considered as a reversible process, the polymer adsorption is assumed to be, mostly, an irreversible process. The isothermal adsorption is generally nonlinear to the polymer concentration, and it can be described by the Freundlich or Langmuir models. The general form of the Freundlich isotherm model is described in Eq. (4.8). The general form of the Langmuir isotherm model is shown in Eq. (4.9). Gupta and Greenkorn (1974) reported the

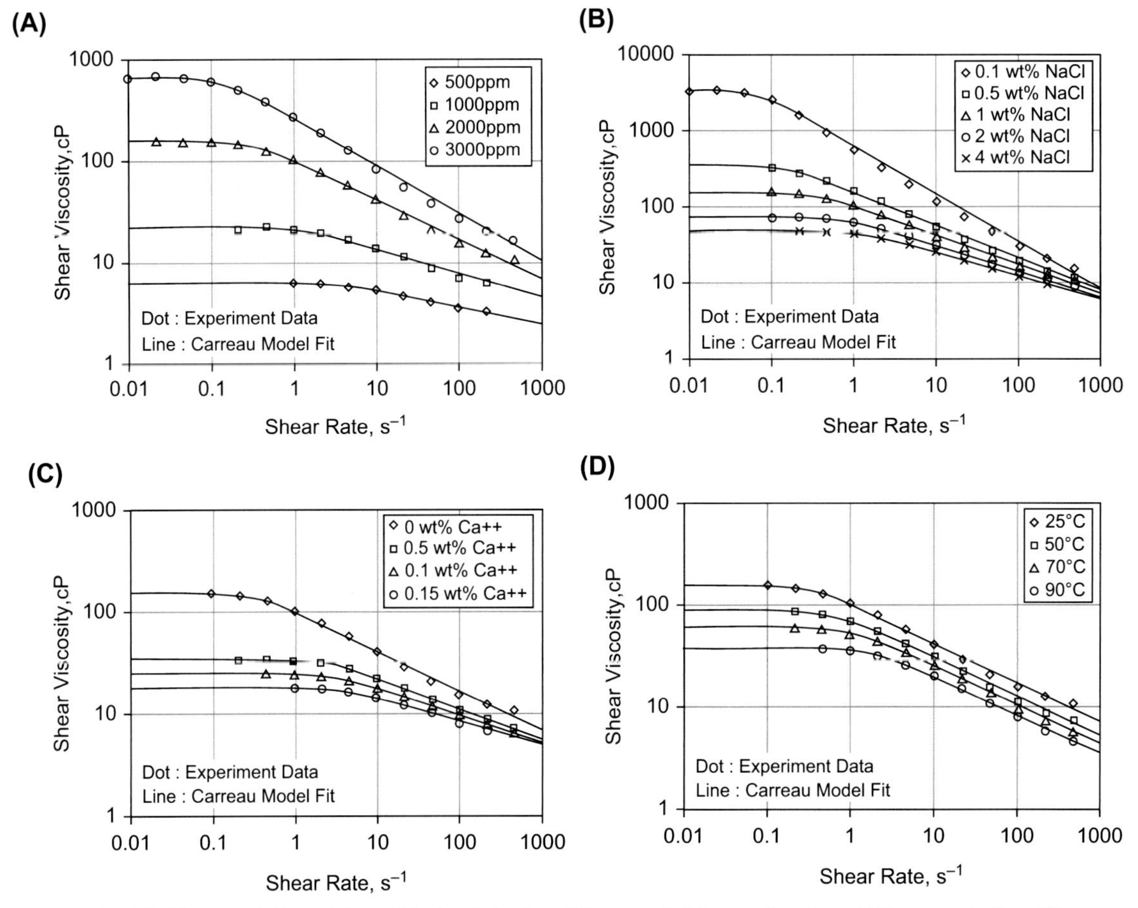

FIG. 4.1 The description of shear-thinning behavior of Flopaam 3630s as a function of **(A)** concentration, **(B)** salinity, **(C)** hardness, and **(D)** temperature. (Credit: From Lee, S., Kim, D. H., Huh, C., & Pope, G. A. (2009). Development of a comprehensive rheological property database for EOR polymers. *Paper presented at the SPE annual technical conference and exhibition, New Orleans, Louisiana, 4–7 October*. https://doi.org/10.2118/124798-MS.)

experimental works to determine the adsorption parameters of the non-linear Freundlich model during the transport of ions in clay-containing porous media.

$$\widehat{C}_i = K_F C_i^n \tag{4.8}$$

$$\widehat{C}_i = \frac{aC_i}{1 + bC_i} \tag{4.9}$$

where $\widehat{C}_i$ is the adsorbed concentration of species i; C_i is the equilibrium concentration of species i in the system; K_F and n are the empirical constants of Freundlich isotherm obtained by fitting experimental data; and a and b are the empirical constants of Langmuir isotherm.

The degree of adsorption is a function of rock surface, salinity, polymer type, molecular weight of polymer, hydrolysis of polymer, and temperature (Sheng, 2011). The rock surface of calcium carbonate has stronger affinity to the carboxylate groups of HPAM compared with the rock surface of silica. The higher salinity in brine increases the level of polymer adsorption. Because the divalent cations contract the size of the flexible HPAM molecules and reduce electrostatic repulsion between the carboxylic group of HPAM and silica surface, the addition of cation promotes the adsorption of HPAM on silica surface. The degree of adsorption varies depending on the polymer type, i.e., nonionic, anionic, and cationic polymer, and molecular weight of polymer. The adsorption of HPAM decreases with an increase in hydrolysis, but there is an optimum hydrolysis for the minimum adsorption. The

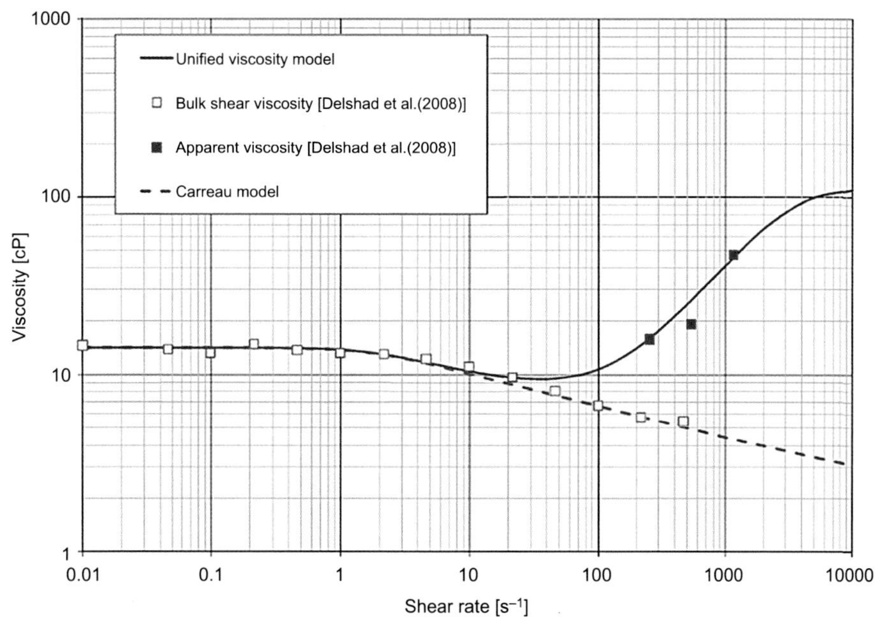

FIG. 4.2 The description of viscoelastic behavior of Flopaam 3630s as a function of shear rate. (Credit: From Kim, D. H., Lee, S., Ahn, C. H., Huh, C., & Pope, G. A. (2010). Development of a viscoelastic property database for EOR polymers. *Paper presented at the SPE improved oil recovery Symposium, Tulsa, Oklahoma, USA, 24–28 April*. https://doi.org/10.2118/129971-MS.)

adsorptions of nonionic and anionic polymers decrease with an increase in the temperature. The mechanism of temperature-dependent adsorption of nonionic polymer is different to that of anionic polymer. For the nonionic polymer, the adsorption is mainly attributed to the attraction by the hydrogen bond. Because the increasing temperature breaks up the hydrogen bond, the attraction between nonionic polymer and rock surface becomes weak, resulting in less adsorption. For the anionic polymer, higher temperature increases the negative charge on the rock surface. The increased electrostatic repulsion between the polymer and rock surface results in the less adsorption of anionic polymer.

Permeability reduction

The adsorption of polymer potentially reduces permeability or blocks pore throat. There are a number of terms to describe the permeability reduction (Sheng, 2011): (1) permeability reduction factor (R_k), which is the ratio of permeability when water flows to the permeability when aqueous polymer solution flows; (2) residual resistance factor (R_{RF}), which is the ratio of water mobility before to after polymer flood; and (3) resistance factor (R_F), which is the ratio of polymer mobility to the water mobility during polymer flow. The cross-linked polymers and gels exploit the permeability reduction to shut off water channeling through high-permeable zones and water conning from the adjacent aquifers. The permeability reduction by the polymer adsorption mainly occurs in water relative permeability rather than oil relative permeability. It is known as the mechanism of the disproportionate permeability reduction (DPR). There are a number of gels: movable gels, pH-sensitive polymer, BrightWater, Microball, preformed particle gel (PPG), etc.

Low Salinity–Augmented Polymer Flood/Gel Treatment

Experiments: polymer flood

Ayirala et al. (2010) developed the desalination scheme for Designer Water developed by Shell. The desalination scheme is designed to formulate the desired injection water composition to suit the formation condition. Using the desalination scheme, the conventional polymer flood can be advanced through incorporating the low-salinity water as makeup brine. The advanced polymer flood using the low-salinity water is the hybrid low salinity–augmented polymer flood (LSPF). In the LSPF process, the cost-competitive LSWF compared with the other EOR methods can bring a couple of benefits to the expensive polymer flood. The study investigated the benefits when the polymer flood exploits the low-salinity water as the makeup brine and evaluated the benefits in terms of the overall economics. In comparison with the conventional polymer flood, which uses the seawater as the makeup brine, the hybrid LSPF introduces a number of advantages. The primary advantage underlying LSPF is the synergetic effect improving both wettability and mobility ratio. As a result, the LSPF improves both volumetric sweep and displacement efficiencies and enhances oil production. The deployment of LSPF is also enabled to avoid a clay swelling and mitigate critical issues, i.e., reservoir souring. In addition, it expects to consume less amount of dosing polymers required to achieve a target viscosity of displacing fluid because of the inherent salinity-dependent viscosity of polymer. The lower concentration of salinity and hardness in LSPF prevent the significant chemical degradation of polymer. The amount of polymer consumption determines the chemical procurement, transportation, storage, and mixing and hydration requirements, and operating costs in offshore environments. The less polymer consumption requires the smaller facilities and reduces the capital expenditures (CAPEX) as well as operating expense (OPEX). The saving cost in the CAPEX and OPEX compensates the extra desalination costs.

The study simulated the analysis of LSPF with the HPAM polymer, Flopaam 3630s. The commercial polymer is most widely used for EOR applications and approximately has 30% hydrolysis and 18–20 million molecular weight. The viscosity of the polymer is experimentally measured at the various temperature and salinity conditions. The experimental results estimate that the polymer concentration requirement achieving the specific target viscosity will be reduced by a factor of 10 when the makeup brine is switched from seawater (35,178 ppm TDS) to the low-salinity designer water (650 ppm TDS). In addition, the study simulated the hypothetical eight designs for LSPF process considering the injection capacity, viscosity of polymeric solution, and existence of desalination process and investigated the CAPEX, space, weight, and OPEX operating LSPF process. The operating costs for the desalination are assumed to be 6% of the total capital costs. The analysis indicates that LSPF requires the higher capital cost by a factor of three and lower OPEX costs per year by a factor of three compared with the conventional polymer flood. The analysis of the present value with 7% discount rate indicates that the payout time corresponding to the extra desalination cost is less than 4 years.

Shaker Shiran and Skauge (2013) experimentally carried out coreflooding using strongly water-wet and intermediate-wet Berea sandstones. The study mainly investigated the efficiency of secondary or tertiary injection modes of LSWF and performance of LSPF on EOR potential. The synthetic seawater of 36,000 ppm TDS is used for the connate water and injecting brine of conventional waterflood. The low-salinity water is prepared by diluting the synthetic seawater by a factor of 10. The commercial Flopaam 3630s is used for polymer injection. The target oil is the diluted crude oil with 2.4 cp, which is favorable for the coreflooding experiment. Because the synthetic seawater and low-salinity water approximately have a viscosity of only 1 cp, the mobility ratio of the conventional waterflood is determined to be unfavorable. The coreflooding of LSWF as secondary or tertiary modes indicates that the early deployment of LSWF is beneficial for the increasing oil recovery. The tertiary LSWF hardly produces the additional oil from the strongly water-wet cores and only recovers the limited oil from the intermediate-wet cores. However, the secondary LSWF is promising to enhance oil production from the both water-wet and intermediate-wet cores. In addition, more enhancing oil production is observed in intermediate-wet cores over water-wet cores. Although these observations are contrast to other experimental observations (Ashraf, Hadia, Torsaeter, & Medad Twimukye Tweheyo, 2010; Rivet, Lake, & Pope, 2010), they clarify the potential of LSWF to modify the wettability of Berea sandstone core. The experiments analyze the performance of LSPF using polymer or linked polymer, i.e., gel, after tertiary LSWF. In the strongly water-wet cores, the applications of LSPF using linked polymer or polymer of 300 ppm increase the pressure differential without producing any additional oil. For the intermediate water-wet cores, the additional oil recovery up to 5% is obtained for the LSPF using the linked polymer. Interesting observation is that LSPF with the linked polymer of 300 ppm shows higher EOR potential than LSPF with the linked polymer of 1000 ppm. It indicates that more favorable mobility ratio is a necessary, but not sufficient conditions for EOR. A following experiment designs the successive deployment of LSWF, conventional polymer flood, and LSPF in the intermediate water-wet cores (Fig. 4.3). The polymer flood after LSWF shows an encouraging response. Because the secondary injection of LSWF ahead of polymer flood already establishes the favorable condition for polymer flood, the polymer flood produces additional oil recovery of 12%. The significant enhanced oil recovery by polymer flood after LSWF results in the negligible contribution of LSPF after the polymer flood on oil recovery. Additional experiments validate the EOR potential of combined injections. They confirm

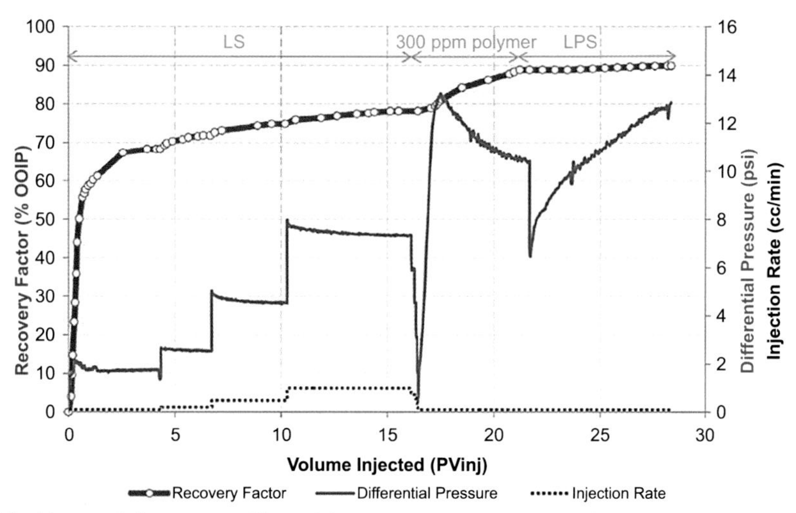

FIG. 4.3 The history of oil recovery, differential pressure, and injection rate from the experiment. (Credit: From Shaker Shiran, B., & Skauge, A. (2013). Enhanced oil recovery (EOR) by combined low salinity water/polymer flooding. *Energy and Fuels, 27*(3), 1223–1235. https://doi.org/10.1021/ef301538e.)

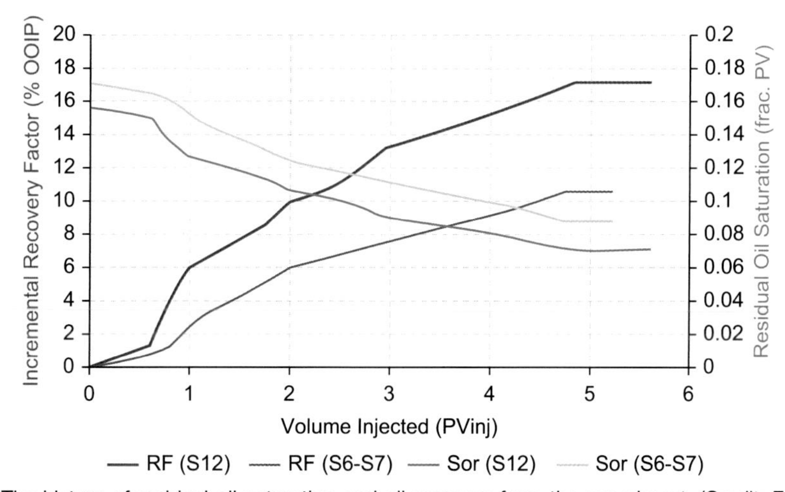

FIG. 4.4 The history of residual oil saturation and oil recovery from the experiment. (Credit: From Shaker Shiran, B., & Skauge, A. (2013). Enhanced oil recovery (EOR) by combined low salinity water/polymer flooding. *Energy and Fuels, 27*(3), 1223–1235. https://doi.org/10.1021/ef301538e.)

the significant reduction in residual oil saturation through the combined process of polymer flood and LSWF (Fig. 4.4). This study experimentally demonstrated the high performance of the combination of LSWF and polymer on the viscous oil recovery. Another study (Shiran & Skauge, 2014) made an effort to investigate the role of the linked polymer on the performance of LSPF.

Shiran and Skauge (2014) investigated how the hybrid LSPF using linked polymer, i.e., low salinity–augmented gel treatment, is different from or similar to the LSPF using polymer in terms of transport in the intermediate water-wet Berea and Bentheimer cores. There are a number of gel types such as strong gel (bulk gel), weak gel (colloidal dispersion gel), or linked polymer solution. In this study, the linked polymer solution of nanosized particles is used. The linked polymer solution normally consists of high-molecular-weight polymer such as HPAM and a cross-linker such as polyvalent cations. It is reported that the gel formation reaction is slow and requires time on the order of days and weeks. This study prepared the linked polymer

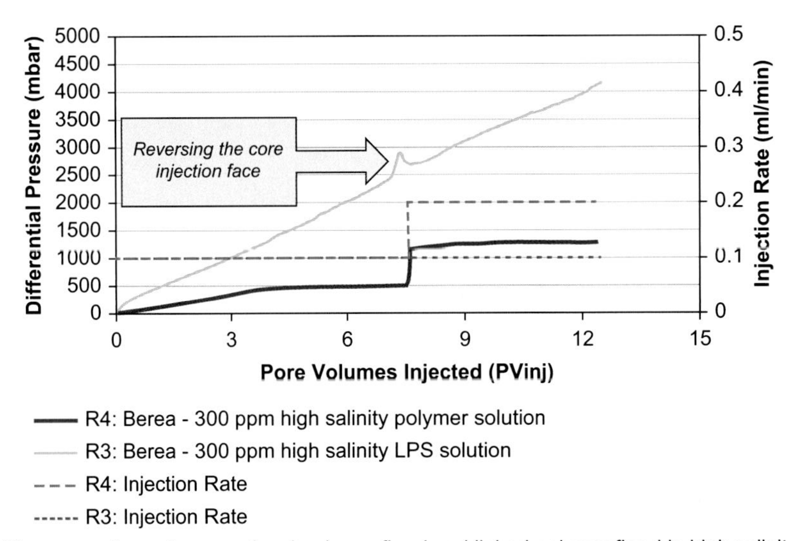

FIG. 4.5 The comparison of conventional polymer flood and linked polymer flood in high salinity condition. (Credit: From Shiran, B. S., & Skauge, A. (2014). Similarities and differences of low salinity polymer and low salinity LPS (linked polymer solutions) for enhanced oil recovery. *Journal of Dispersion Science and Technology*, *35*(12), 1656–1664. https://doi.org/10.1080/01932691.2013.879532.)

solution using Flopaam 3630s and aluminum citrate as cross-linker. The polymer of 300 ppm is used for both LSPF and gel treatment.

The study conducted the coreflooding of single-phase and two-phase displacements. Firstly, the displacement experiments of the single phase are carried out to compare the propagation of polymer and linked polymer solution at different salinity conditions. The change of differential pressure corresponding to the injection rate is measured to quantify the transport of polymer and linked polymer in the porous media. In the single-phase experiments using the low-permeable Berea sandstone, the residual resistance factors between LSPF and low salinity–augmented gel treatment are compared. The higher residual resistance factor with 7% is observed in low salinity–augmented gel treatment than the LSPF. However, the difference is acceptable considering the initial permeability difference. It indicates that the propagation behavior between gel treatment and polymer flood is approximately similar in the low saline, low permeable, and intermediate water-wet conditions. For the high-permeable Bentheimer cores, both the low salinity–augmented gel treatment and LSPF processes show the identical history of differential pressure with injection rate. These observations also confirm the comparable transport of both polymer and linked polymer in low saline, high permeable, and intermediate water-wet conditions. Additional single-phase experiments monitor the differential pressure of conventional polymer flood and gel treatment using synthetic seawater in high saline condition (Fig. 4.5). It is observed that the differential pressure of polymer flood increases by the significant adsorption of polymer until the adsorption of polymer reaches to the maximum adsorption level. For the experimental results of gel treatment, continuously increasing differential pressure is observed. It is understood that the excess amount of divalent cations bridges the negatively charged polymer molecules and results in the more aggregation of linked polymer. The size of aggregation in high salinity condition becomes larger than pore throat and introduces the higher mechanical entrapment. In comparison with the results of high salinity conditions, the adsorption of polymer on to the rock surface and the arrogation of linked polymer can be reduced in the low salinity condition. The propagations of polymer and linked polymer become to be similar in the low salinity condition. The second experiments of two-phase displacement investigate the change of residual oil saturation of LSPF and low salinity–augmented gel treatment following secondary LSWF. It is observed that the reduction of residual oil saturation up to 48.8% and 12.9% by LSPF and low salinity–augmented gel treatment, respectively, in the intermediate water-wet cores, but negligible change in the fully water-wet cores. Although these observations show the higher performance of LSPF without cross-linker than that with cross-linker in one-dimensional and homogeneous system, the heterogeneity of system can provide different results.

Vermolen, Pingo Almada, Wassing, Ligthelm, and Masalmeh (2014) validated the prospective benefits and synergetic effects, which can be drawn for LSPF. It is summarized that the low salinity makeup brine can secure a number of benefits to polymer flood: (1) chemical stability; (2) mechanical stability, especially, at high temperature condition; (3) less adsorption; and (4) less scaling and souring problems. The benefits provide less amount of polymer required to reach the target viscosity as observed in previous studies. In addition, it is suggested that the LSPF introduces the mechanism of LSWF and using low-salinity water of makeup brine enhances the elasticity of polymer. The mechanism of LSWF confidentially modifies the wettability of reservoir. In addition, the elastic behavior of polymer potentially reduces the residual oil saturation. Because of the wettability modification and more reduction in residual oil reduction, the LSPF can enhance the displacement efficiency compared with the conventional polymer flood. This study tried to demonstrate the benefits by measuring the polymer behavior under various conditions. The first experiment of the study measures the amounts of HPAM polymer concentration to reach the target viscosity of polymeric solution at the different salinity conditions of 7000 and 700 ppm TDS. About half amount of polymer is required to reach the target viscosity with 50 cp at a shear rate of 11.5 s^{-1} in low salinity compared with the high salinity conditions. Another experiment estimates 34 times reduction in polymer concentration at the salinity of 1500 ppm TDS compared with the 260,000 ppm TDS. Because the extra cost in OPEX/CAPEX for the desalination is much lower than the saving in OPEX for the less polymer injection, securing chemical stability of polymer using low-salinity makeup water significantly increases the economics of LSPF. The second experiment evaluates the viscosity loss by the mechanical degradation at different shear rate and salinity conditions. The experiment calculates the degree of polymer degradation, and it is observed that the viscosity loss by shear force increases with an increase in salinity as shown in Fig. 4.6. The third experiment using dynamic (oscillatory) frequency sweep measurements analyzes the viscoelasticity of polymer varying polymer concentration at different salinity conditions. It measures the relation between storage (elastic) modulus and loss (viscous) modulus and determines the relaxation time of the polymer from the relation. The higher relaxation time is observed with decreasing salinity, which means the increasing viscoelastic characteristics of polymer. The increasing viscoelastic characteristics of polymer mobilize the immobile oil and potentially increase the oil recovery. The polymer adsorption during LSPF is also briefly investigated in two aspects. Firstly, the low-salinity water as makeup brine leads to the less adsorption of polymer and the degree of the reduction varies with polymer and rock types. Secondly, the effect of mixing between displacing and displaced brines on the adsorption is investigated. When the low-salinity polymeric solution displaces high-salinity connate water, polymer adsorption occurs and polymer propagation is retarded. Until the polymer adsorption reaches

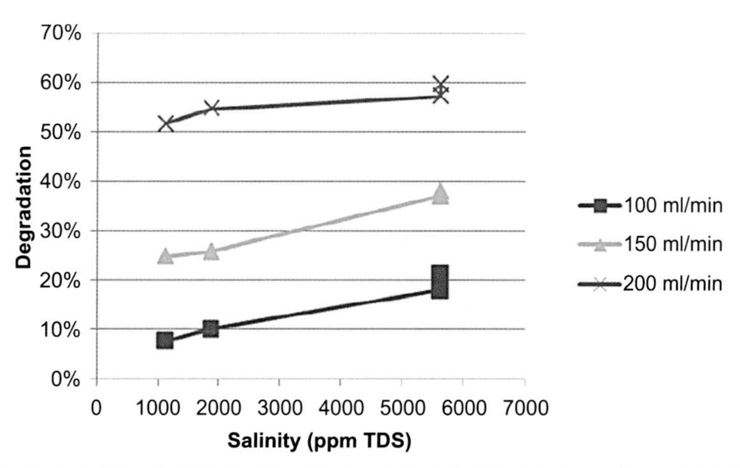

FIG. 4.6 Shear degradation of viscosity of polymeric solution at different share rates and salinity conditions. (Credit: From Vermolen, E. C. M., Pingo Almada, M., Wassing, B. M., Ligthelm, D. J., & Masalmeh, S. K. (2014). Low-salinity polymer flooding: Improving polymer flooding technical Feasibility and economics by using low-salinity make-up brine. Paper presented at the International petroleum technology conference, Doha, Qatar, 19–22 January. https://doi.org/10.2523/IPTC-17342-MS.)

to the maximum adsorption capacity, the buffer zone of low-salinity water forms behind the connate water and the head of polymer front in porous media. Although the buffer zone can protect the rheology of low-salinity polymeric solution from the interference by high-salinity connate water, it delays the oil recovery.

Although the study addressed the promising benefits of LSPF, it also cautioned a couple of potential risks to be involved in LSPF. The usage of low-salinity water during LSPF disturbs the original equilibrium state in the reservoir and causes the reequilibrium. During the reequilibrium by low-salinity water intrusion, the cation exchange of the clay and low salinity in the bulk solution might destabilize the clay and lead to more clay swelling by a double layer expansion. In addition, the reequilibrium might involve the mineral dissolution, which increases the multivalent cations in the solution causing chemical degradation of polymer. The electrostatic interaction between the polyelectrolyte polymer and the clay surface also can be affected by the reequilibrium process. It is concluded that the evaluations of the positive benefits and negative risks are required to do a full economics evaluation of LSPF and optimization to improve the economic viability of the hybrid process.

Previous studies have observed the increasing oil production by the synergy of LSPF and measured the improving polymer rheology and stability by using the low-salinity water as makeup brine. However, they have not demonstrated whether the LSPF still secures the mechanism of LSWF or not. They only explained that the higher recovery of LSPF than polymer flood or LSWF is the evidence of securing the mechanism of LSWF with the assistances from the improving rheology and stability of polymer.

A couple of studies (AlSofi et al., 2016, AlSofi, Wang, & AlBoqmi, 2018) have evaluated the synergy between chemical EOR and smart waterflood and evaluated the possibility of the hybrid EOR on the oil recovery increase in a more realistic framework. The studies have investigated both the polymer flood and surfactant flood as chemical EOR candidates. Experiments with polymer flood are only illustrated herein and those with surfactant flood will be discussed later. The studies tried to clarify the possible synergy by performing a suite of experiments including electrokinetics potential, contact angle, rheological, coreflood, and NMR tests. The target reservoir is the slightly viscous Arabian heavy carbonate reservoir and has the high temperature and high salinity conditions. The temperature is about 99°C, and connate water has the salinity of about 244,000 mg/L and hardness of 58,000 mg/L (as $CaCO_3$). The injecting brine has the salinity of about 69,000 mg/L and the hardness of 9000 mg/L. The smart water or low-salinity water is prepared by the dilution of the injecting brine by a factor of 10. The core and dead crude oil for the experiments are obtained from the target reservoir. The dead crude oil has viscosity of about 4.39 cp at the reservoir condition. Because the crude oil has gas-oil ratio less than 100 scf/bbl, the viscosity of crude oil is close to that of dead oil. Consequently, the experiments using the dead oil accurately represent that using live oil.

In the study, the first experiment measures the rheology of sulfonated polyacrylamide polymer at three temperatures of 25, 40, and 60°C. The viscosity of polymeric solution is measured as polymer concentration and brine type of makeup brine change. Because the reservoir temperature is higher than the experimental temperatures, the viscosity model of power law is constructed using the experimental rheology measurements and the viscosity of polymeric solution at the reservoir temperature is estimated. For a target viscosity of 11 cp, it is determined that 30% of polymer concentration can be reduced when the brine type is switched from the injecting water to low-salinity water. The injection scheme of coreflooding is designed considering the result of rheology experiments. The second experiment using PALS technique measures an electrophoretic mobility and estimates a surface electrokinetics potential, i.e., ζ-potential, in the various binary systems of a crushed reservoir rock sample and a variety of polymeric solutions. The polymeric solutions are prepared at different concentrations of polymer and different brine types. It is, conventionally, known that successful LSWF decreases the potential toward negative of ζ-potential. Firstly, the experiments measure the electrophoretic mobility and calculate the ζ-potentials of the high-salinity injecting brine and low-salinity water without polymer (Fig. 4.7). Both the mobility and potential show positive values for injecting brine and negative values for low-salinity water. It is clearly demonstrated that the LSWF possibly introduces the wettability modification compared with the conventional waterflood. The various polymeric solutions, using the injecting brine or low-salinity water as makeup brine, are investigated (Fig. 4.8). It is observed that the presence of polymer shifts the electrophoretic mobility from positive to negative values for the high-salinity injecting brine. Because the polymer has the anionic charges on the backbone and a group of counterions in the bulk solution can form an electrical double layer around the charged component in polymer, the effects of the potential-determining ions might be shielded

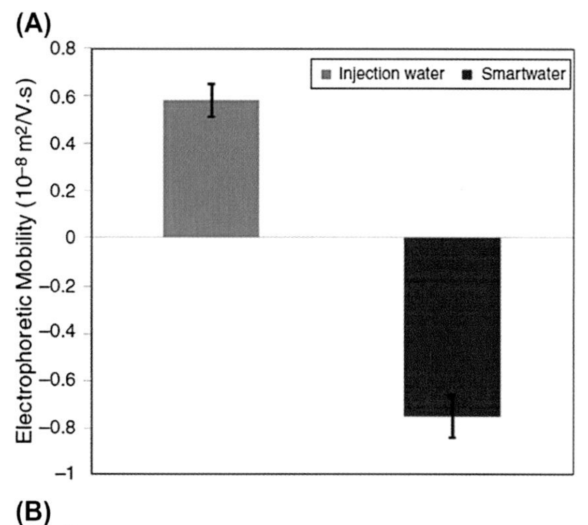

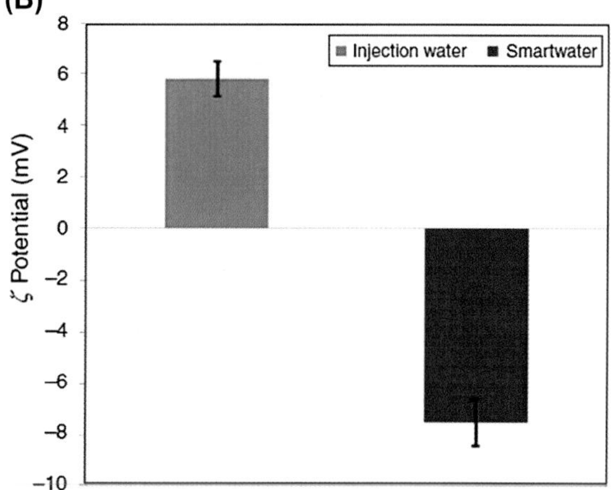

FIG. 4.7 Comparison of **(A)** electrophoretic mobility and **(B)** ζ-potential between high-salinity injecting brine and low-salinity water. (Credit: From AlSofi, A. M., Wang, J., AlBoqmi, A. M., AlOtaibi, M. B., Ayirala, S. C., & AlYousef, A. A. (2018a). Smartwater synergy with chemical enhanced oil recovery: Polymer effects on Smartwater. *SPE Reservoir Evaluation & Engineering Preprint (Preprint)*, 17. https://doi.org/10.2118/184163-PA.)

by the charged polymer and effective concentrations of potential-determining ions decrease. Therefore, the polymer addition is not associated with the electrical double layer expansion of LSWF mechanism as well as the wettability modification. Rather than the wettability modification, it indicates the potential of polymer adsorption to the suspended rock particles. For the low-salinity polymeric solution, there is a huge decrease toward negative electrophoretic mobility compared with high-salinity polymeric solution. These observations guarantee that the presence of polymer might

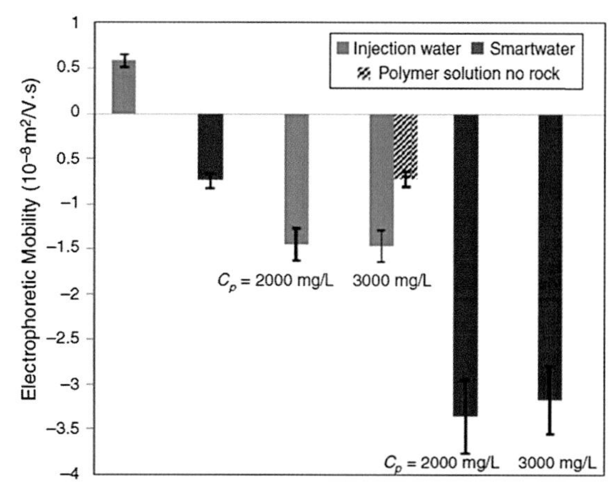

FIG. 4.8 Comparison of electrophoretic mobility of carbonates between brines and polymeric solutions. (Credit: From AlSofi, A. M., Wang, J., AlBoqmi, A. M., AlOtaibi, M. B., Ayirala, S. C., & AlYousef, A. A. (2018a). Smartwater synergy with chemical enhanced oil recovery: Polymer effects on Smartwater. *SPE Reservoir Evaluation & Engineering Preprint (Preprint)*, 17. https://doi.org/10.2118/184163-PA.)

slightly enhance or not hinder the mechanism of LSWF for carbonate reservoir. The third experiment investigates the contact angle in the system of oil/polymeric solution/calcite disk or reservoir rock. Preliminary study measures the contact angles for injecting water and low-salinity water without polymer addition and observes the reduction of contact angle when the salinity of brine reduces. In the measurements of polymeric solution, the addition of polymer reduces the contact angle for the reservoir rock sample but provides negligible change for the calcite disk. Lastly, experimental coreflooding simulates the deployments of conventional polymer flood using high-salinity injecting brine and LSPF. Following the previous results from rheology measurement, the conventional polymer flood incorporates 30% higher concentration of polymer than the LSPF to achieve equivalent target viscosity. Obviously, the tertiary mode of the conventional polymer flood recovers 18.2% additional oil over the secondary conventional waterflood. The following LSPF process increases the oil recovery by 6.5% over the tertiary polymer flood. Another coreflooding test confirms the improving oil recovery of LSPF compared with the polymer flood and LSWF. The NMR test also observes the change in connectivity of pores. These studies clearly have quantified the potentials of hybrid LSPF to introduce the wettability modification in carbonate reservoir and improving the rheology of polymeric solution. It is also concluded that the hybrid LSPF is effective to recover heavy oil from carbonate reservoirs.

As well as the benefits and synergies underlying LSPF, it is necessary to investigate another potential risk of LSPF. Another study (AlSofi, Wang, & Kaidar, 2018) from the same research group analyzed the injectivity and polymer retention of LSPF at the dynamic condition. The ionic composition of low-salinity water potentially expands the adsorbed polymer molecular and controls the level of retention (Fig. 4.9). The injectivity is the crucial factor at wellbore region and highly influences the injecting capacity of polymer EOR process. The retention is also related to the polymer loss and unexpected transport of polymer. Therefore, this study evaluated the injectivity, inaccessible pore volume, and retention of hybrid LSPF compared with the conventional polymer flood. The experimental system of the study is equal to that of previous studies (AlSofi et al., 2016; AlSofi, Wang, & AlBoqmi, 2018), neglecting the heavy oil flow. Because the study made an effort on investigating the flow behavior of polymer at the

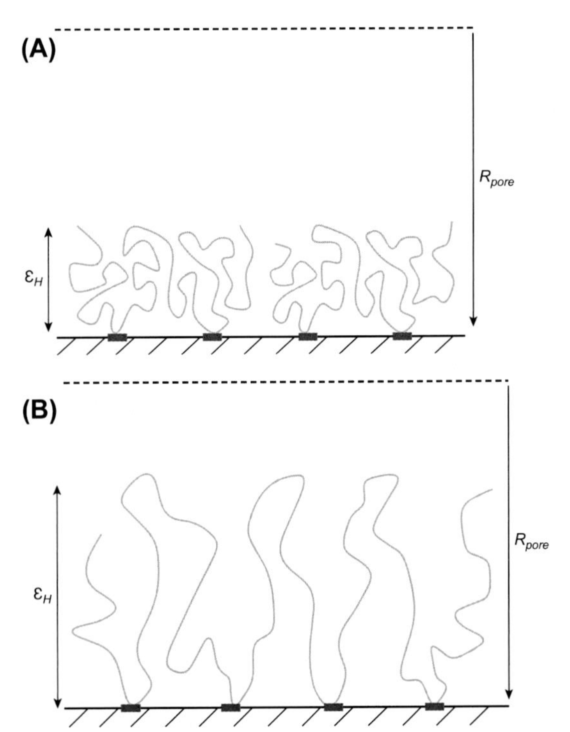

FIG. 4.9 Thickness of adsorbed polymer molecular at **(A)** high-salinity water condition and **(B)** low-salinity water condition. (Credit: From AlSofi, A. M., Wang, J., & Kaidar, Z. F. (2018b). SmartWater synergy with chemical EOR: Effects on polymer injectivity, retention and acceleration. *Journal of Petroleum Science and Engineering, 166*, 274–282. https://doi.org/10.1016/j.petrol.2018.02.036.)

different salinity conditions, it carried out single-aqueous phase displacement experiment without considering oleic phase. The same polymeric solutions of the conventional polymer flood and LSPF are used as previous studies (AlSofi et al., 2016; AlSofi, Wang, & AlBoqmi, 2018). The single-phase coreflooding is designed with two cycles of low salinity and high salinity conditions. Each cycle is composed of the three injection phases: water, polymeric solution, and water. With the injection design, the experiment consists of the four sets of coreflooding: two sets for injectivity analysis and additional two sets for polymer retention and/or acceleration analysis. Measuring the differential pressure with the flow rate, injectivity of the process is calculated for the two coreflooding tests. Incorporating the total organic carbon (TOC) measurements and gas chromatography (GC) measuring polymer and tracer concentrations, the effluent concentration data from the additional two coreflooding tests are analyzed to estimate the polymer retention/acceleration and inaccessible pore volume.

The injectivity test of polymer flood determines the resistance factor (R_F) and residual resistance factor (R_{RF}) to analyze the injectivity quantitatively. At the high injection rate indicating near wellbore region condition, the usage of low-salinity water for makeup brine introduces the negative impact on the injectivity of polymer and chase water increasing both resistance factor and residual resistance factor. In addition, the relation of Eq. (4.10) between the factors calculates the in situ viscosity of conventional polymer flood and LSPF to describe the in situ rheology of polymeric solution.

$$\eta_{in-situ} = \mu_w \frac{R_F}{R_{RF}} \tag{4.10}$$

where $\eta_{in-situ}$ indicates the in situ viscosity of polymeric solution.

Switching the flow rate and brine type, the in situ viscosity changes as shown in Fig. 4.10. The variation in the in situ viscosity implies that all polymeric solutions using high-salinity injecting water or low-salinity water suffer shear-thickening behavior, which shows an increasing viscosity with an increase in shear rate. An interesting observation is less in situ viscosity for the low-salinity polymeric solution compared with the high-salinity polymeric solution at the same shear rate. Conventionally, the apparent viscosity of polymeric solution should increase at the low salinity condition improving chemical stability. However, the experimental results show the opposite trend despite the higher chemical stability of polymer. This opposite

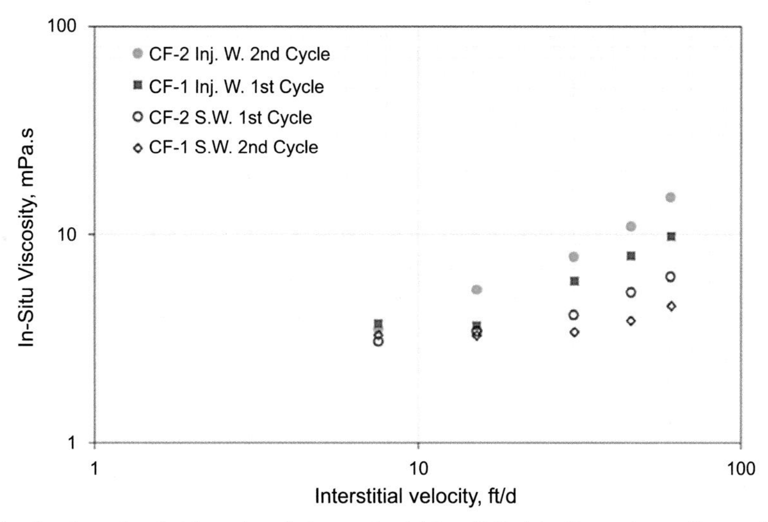

FIG. 4.10 In situ viscosity of polymeric solutions using high-salinity injecting water and low-salinity water at various shear rates. (Credit: From AlSofi, A. M., Wang, J., & Kaidar, Z. F. (2018b). SmartWater synergy with chemical EOR: Effects on polymer injectivity, retention and acceleration. *Journal of Petroleum Science and Engineering*, *166*, 274–282. https://doi.org/10.1016/j.petrol.2018.02.036.)

result is explained that the low-salinity polymeric solution exhibits the significant shear-thinning behavior before the shear-thickening behavior. The second experiment estimates the polymer retention at dynamic condition using the polymer mass balance equation of Eq. (4.11).

$$\Gamma_{\text{dynamic, poly}} = \frac{V_{\text{inj}} C_{\text{inj, poly}} - \sum_{n} V_{\text{prod}} C_{\text{prod, poly}}}{W_{\text{rock}}} \tag{4.11}$$

where $\Gamma_{\text{dynamic, poly}}$ is the polymer retention at dynamic condition; V_{inj} is the injected volume of polymeric solution; $C_{\text{inj, poly}}$ is the polymer concentration to be injected; V_{prod} is the produced volume of polymeric solution; $C_{\text{prod, poly}}$ is the polymer concentration to be produced; and W_{rock} is the weight of the rock.

In the two sets of coreflooding, the conventional polymer flood results in the polymer dynamic retention of 0.230 and 0.133 mg/g of rock. The LSPF yields the retention of 0.084 and 0.102 mg/g. The usage of low-salinity water reduces the 1028% of reduction in the dynamic condition and will save the costs mitigating polymer loss in the bulk of polymeric solution. In addition, the experiments also estimate the inaccessible pore volume by polymer molecules, which differentiates the depletion in the effluent profile of polymer from that of tracer. The inaccessible pore volume, in which polymer hardly enters, might promote the faster flow of polymeric solution and contribute the acceleration of the polymer front. The coreflooding tests of conventional polymer flood have the inaccessible pore volumes of 0.136 and 0.142. LSPF shows the higher inaccessible pore volume with 0.147 and 0.149. The slight increase in the inaccessible pore volume is attributed to the expansion of polymer molecule in low salinity condition. The increment in the inaccessible pore volume might improve or, at least, not hinder the propagation of polymer front. Because the increment is under the range of experimental uncertainty, the experimental results might not guarantee the favorable effect of inaccessible pore volume on the flow of polymer in porous media. It is concluded that the successful LSPF should be carefully deployed considering the issues of injectivity, dynamic retention, and inaccessible pore volume and then synergy of LSPF might produce successful oil recovery.

For the heavy oil sandstone reservoir, Almansour, AlQuraishi, AlHussinan, and AlYami (2017) also carried out the lab-scaled experiments to investigate the potential of LSPF process. The study measured the IFT, contact angle, and ζ-potential to quantify the wettability modification effect, but it tested the brines, not polymeric solution. The heavy oil recovery of displacement experiment and rheology of polymeric solution are also investigated. The Saudi heavy crude oil is subject to the experiments and has viscosity of 33 cp and density of 0.91 g/cm^3 at 60°C and 2000 psi. The synthetic formation water of 197,451 ppm TDS is prepared. The Arabian Gulf seawater of 36,170 ppm TDS and the diluted version of the seawater by a factor of 10, i.e., low-salinity water, are used. The following experiments investigate the cores from the Berea and low clay content Bentheimer sandstones. The polymeric solution of LSPF is prepared with HPAM of 5000 ppm and low-salinity water.

Firstly, IFT measurements between two brines and heavy oil clearly provide the reduction of IFT up to 20 dyne/cm as the salinity decreases. It is explained that the reduction is not enough to modify wettability and increase oil production. This conclusion agrees to the extensive studies of LSWF. Varying the salinity of brine, the contact angles in Bentheimer and Berea sandstones are measured. In the formation water and seawater conditions, the wettability of Berea sandstone is determined as intermediate wetness with contact angle of 90 degrees. Switching the brine with low-salinity water drops contact angle down to 75 degrees, which indicates the wettability modification of Berea sandstone. For the low clay content Bentheimer sandstone core, the contact angle of 70°C is measured in the formation water and seawater conditions and the wettability is determined to be close to water-wetness. The low-salinity water only decreases the contact angle by 5 degrees; therefore, it hardly modifies the wettability of Bentheimer core. The ζ-potential measurement draws the similar conclusions, which are obtained from the contact angle measurement. The seawater results in the positive value of ζ-potential, and the significant reduction in ζ-potential is obtained for the low-salinity water. The low-salinity water also decreases the ζ-potential of Bentheimer core compared with the seawater. The magnitude of the ζ-potential reduction in Bentheimer core is highly less than that in Berea core. A number of corefloodings are designed as secondary mode waterflood or LSWF and tertiary mode of conventional polymer flood or LSPF. Using the intermediate-wet Berea sandstone core, it is observed that the secondary LSWF shows higher oil recovery than the secondary waterflood. The tertiary mode of LSPF also enhances the oil recovery over the tertiary mode of conventional polymer flood or secondary mode of LSWF. It is obvious that the wettability modification mechanism during LSWF and LSPF is effective in intermediate-wet Berea sandstone. In contrast, the effect of wettability modification is not observed in both processes of secondary LSWF and tertiary LSPF for the water-wet Bentheimer sandstone core. This contrasting observation between Berea and Bentheimer sandstones is comparable with the observations from contact angle and ζ-potential measurements. This study concluded that LSPF enables to improve heavy oil recovery from sandstone core. Initial wettability and clay content are also of importance to employ wettability modification effect.

Torrijos et al. (2018) also investigated the application of LSPF in terms of the initial wetting, wettability alteration with improving microscopic sweep efficiency and redistribution of oil, and mobility of oil within the sandstone core. The stabilized crude oil has AN of 0.10 mg KOH/g and BN of 1.80 mg KOH/g. The formation water has the salinity of 100,000 ppm TDS, seawater has salinity of 33,390 ppm TDS, and low-salinity water consisting of only NaCl has salinity of 1000 ppm TDS. The polymeric solution of LSPF is prepared by dissolving the 1000 ppm of Flopaam 3630s polymer into the low-salinity water of 1000 ppm NaCl brine. Although the brines of formation water and low-salinity water have the low pH of 5.5 and 5.7, the fluids of seawater and low-salinity polymeric solution have the high pH of 7.8. Three sets of sandstone cores are subject to the coreflooding experiments. A majority of the sandstone cores are composed of quartz, albite, and illite. Because the surface area and CEC of sandstone are the crucial factors to determine the degree of low-salinity waterflood effect, the surface area of core materials is measured using Brunauer-Emmett-Teller (BET) surface area measurement. The cores have the equivalent surface area of $1.81 \pm 0.02\ m^2/g$. Before the description of LSPF experiments, the study summarized the previous experimental results of LSWF.

The experiments of imbibition test observe the increasing oil recovery with an increasing pH up to 9 when the injecting brine is switched from secondary formation water to the tertiary low-salinity water. It is explained that initial wettability is the fractional wetness and suggested that the observation is in line with the suggested mechanism of the pH increase modifying wettability. The increasing pH, alkaline condition, by LSWF is favorable to the deprotonation of acidic and basic polar organic components of oil. The interaction of the deprotonated organic components onto the negatively charged clay surface decreases shifting wettability toward more water-wet. Recalling the experiments of tertiary LSWF process, the deployment of hybrid LSPF following the secondary LSWF is experimentally performed. In the experiment, the secondary LSWF process recovers the 66% of OOIP, which shows the 17% higher recovery of OOIP compared with the previous results of tertiary LSWF process. Following the suggested mechanism, the higher oil recovery of secondary LSWF process should use the higher pH condition over tertiary

LSWF process. However, the pH condition less than 7.5 is observed in the secondary LSWF and it is lower than that in the tertiary LSWF. Despite the unexpected history of effluent pH corresponding to the mechanism, it is obviously important to determine the deployment of LSWF as secondary or tertiary mode for successful oil production. Following the secondary LSWF, tertiary LSPF is successively applied. It enhances oil recovery by 22%, which is originated from the improving sweep efficiency (Fig. 4.11). Another experiment evaluates the oil recovery from the secondary LSPF and tertiary LSWF. This experiment observes only 68% of OOIP after secondary LSPF and negligible improvement during tertiary LSWF. In comparison between the two experiments, the secondary LSPF only shows 2% higher recovery than secondary LSWF. The tertiary LSPF produces the additional oil recovery by 18% than secondary LSPF. It is explained that the secondary LSWF is more efficient to modify wettability and redistribute the residual oil within the pore space and tertiary LSPF easily mobilizes the redistributed residual oil as shown in Fig. 4.12.

Unsal, ten Berge, and Wever (2018) also investigated the potential aspects of LSPF through single-phase displacement experiment being similar to the AlSofi, Wang, and Kaidar (2018). Unsal et al. (2018) focused on the LSPF in sandstone reservoirs, not carbonate reservoirs. The study investigated the cation exchange in the presence of polymer as well as injectivity and polymer retention. Because some polymer molecules have the polyelectrolyte characteristic, the presence of polymer could cause exchange cations influencing the clay swelling and compatibility regime between formation and injecting brines. During the single-phase coreflooding, the surface area of porous media is fully subject to the only polymer adsorption, not oil adsorption, and the adsorption in the single-phase system can be overestimated compared with that in the multiphase system. Firstly, the effluent concentration of polymer through single-phase coreflooding is measured to investigate the polymer retention. Two polymeric solutions of conventional polymer flood and LSPF are investigated for the measurements. The two cycles of coreflooding are carried out for each measurement. The first cycle of polymeric solution injection fully contributes to the polymer retention on the core. Therefore, the second cycle of polymeric solution injection into the same core is assumed not to be affected by the retention. To quantify the delay of polymer transport due to the retention, the additional coreflooding tests injecting tracer are carried out and compared with the corefloodings of polymeric solution injection (Fig. 4.13). The difference between the effluent profiles of tracer and polymer corresponds to the delay of polymer production. The coreflooding of conventional polymer flood shows the delay of 0.35 PV (Fig. 4.13A), and the coreflooding of LSPF shows the delay of 0.05 PV (Fig. 4.13B). These results correspond to the late

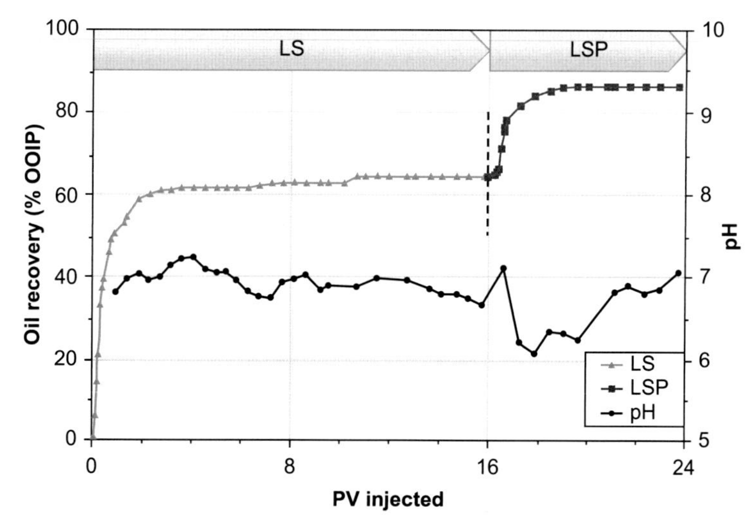

FIG. 4.11 History of oil recovery and pH through secondary low-salinity water flood and tertiary low-salinity polymer flood. (Credit: From Torrijos, P., Iván, D., Puntervold, T., Skule Strand, Austad, T., Bleivik, T. H., et al. (2018). An experimental study of the low salinity smart water – polymer hybrid EOR effect in sandstone material. *Journal of Petroleum Science and Engineering, 164*, 219–229. https://doi.org/10.1016/j.petrol.2018.01.031.)

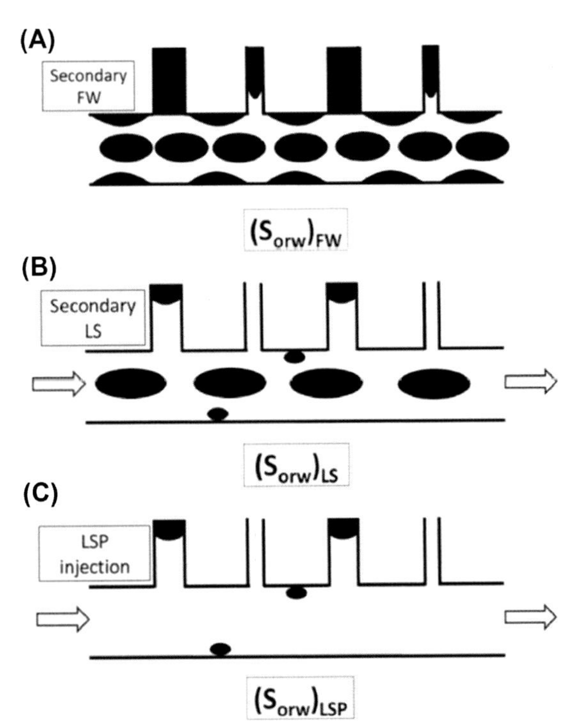

FIG. 4.12 Description of distribution and mobilization of residual oil saturation after **(A)** secondary injection of formation water; **(B)** secondary low-salinity waterflood; and **(C)** tertiary low-salinity polymer flood. (Credit: From Torrijos, P., Iván, D., Puntervold, T., Skule Strand, Austad, T., Bleivik, T. H., et al. (2018). An experimental study of the low salinity smart water – polymer hybrid EOR effect in sandstone material. *Journal of Petroleum Science and Engineering, 164*, 219–229. https://doi.org/10.1016/j.petrol.2018.01.031.)

breakthrough time by 22% and are attributed to higher retention of polymer by a factor of 5 in conventional polymer flood than LSPF. Secondly, the long-term injectivity test monitors the stabilized pressure profiles of conventional polymer flood and LSPF. The LSWF for 150 PV follows the conventional polymer flood for 130 PV. Before the initiation of LSPF, high-salinity and low-salinity brines are flooded to exclude the uncertainty. The test results in the higher differential pressure for conventional polymer flood and lower differential pressure for LSPF, which means the improved injectivity is obtained for LSPF. This observation is contrast with the observation in the carbonate reservoirs (AlSofi, Wang, & Kaidar, 2018). Unsal et al. (2018) explained that the increasing injectivity of polymer flood with low-salinity makeup brine in sandstone is attributed to the role of divalent ions. The higher effluent concentration of divalent cations for LSPF over that for conventional polymer flood is observed. The higher injectivity in LSPF is explained that some polyelectrolyte polymer can form complexes with the divalent ions, and the consumption of divalent ions, in turn, forces to release the additional divalent ions from the rock surface to bulk of polymeric solution for reequilibrium. The formation of complexes enables to explain the less apparent viscosity (Fig. 4.10) and differential pressure. This study described a number of conclusions regarding LSPF: (1) LSPF reduces the delay of polymer propagation owing to less retention and expects the less delay of oil bank arrival; (2) although current experiments show the improved injectivity, general conclusion can be developed with the more extensive experimental observations; (3) formation of complexes by the cation exchange between the hydrolyzed parts of polymer and divalent cations can be another risk to decrease the apparent viscosity of LSPF.

Experiments: Gel treatment

The polymer gels are frequently applied for the conformance control to plug the high conductivity of channels of reservoirs. The plugged polymer in fractures reduces the fracture conductivity and mitigates the early breakthrough. The high injection pressure exceeding the pressure of gel rupture might cause to reopen the fractures. The successful gel treatments incorporate the maintaining of stable rupture pressure of gel. Another key issue for the successful deployment is the gel swelling affecting the fluid flow. The salinity difference between gel network and aqueous phase might influence the degree of swelling. Brattekås, Graue, and Seright (2016) experimentally investigated the effects of the salinity on the blocking performance of conventional polymer gel, Cr(III)-acetate HPAM gels. They constructed the water-wet fractured core plugs and placed the formed gels, which are composed of 5000 ppm of HPAM and 427 ppm of Cr(III)-acetate, in the fracture core plugs. The gel solvent is the formation water of salinity with 79,170 ppm, and the high-salinity formation water is obtained from the North Sea chalk reservoir. In the fractured cores with formed gel, four different saline brines are flooded. Including formation water, the three low-salinity brines have the salinities of 1,000, 500, and 0 ppm as NaCl. The n-decane as oil is subject to the experiment. In the system, the five sets of coreflooding using the brines measure the differential pressure across the matrix and fracture as well as production rate. For the three sets of coreflooding, each test is designed with five cycles of injection, and differential pressure and production rates are measured (Fig. 4.14). Firstly, formation water is injected and low-salinity water

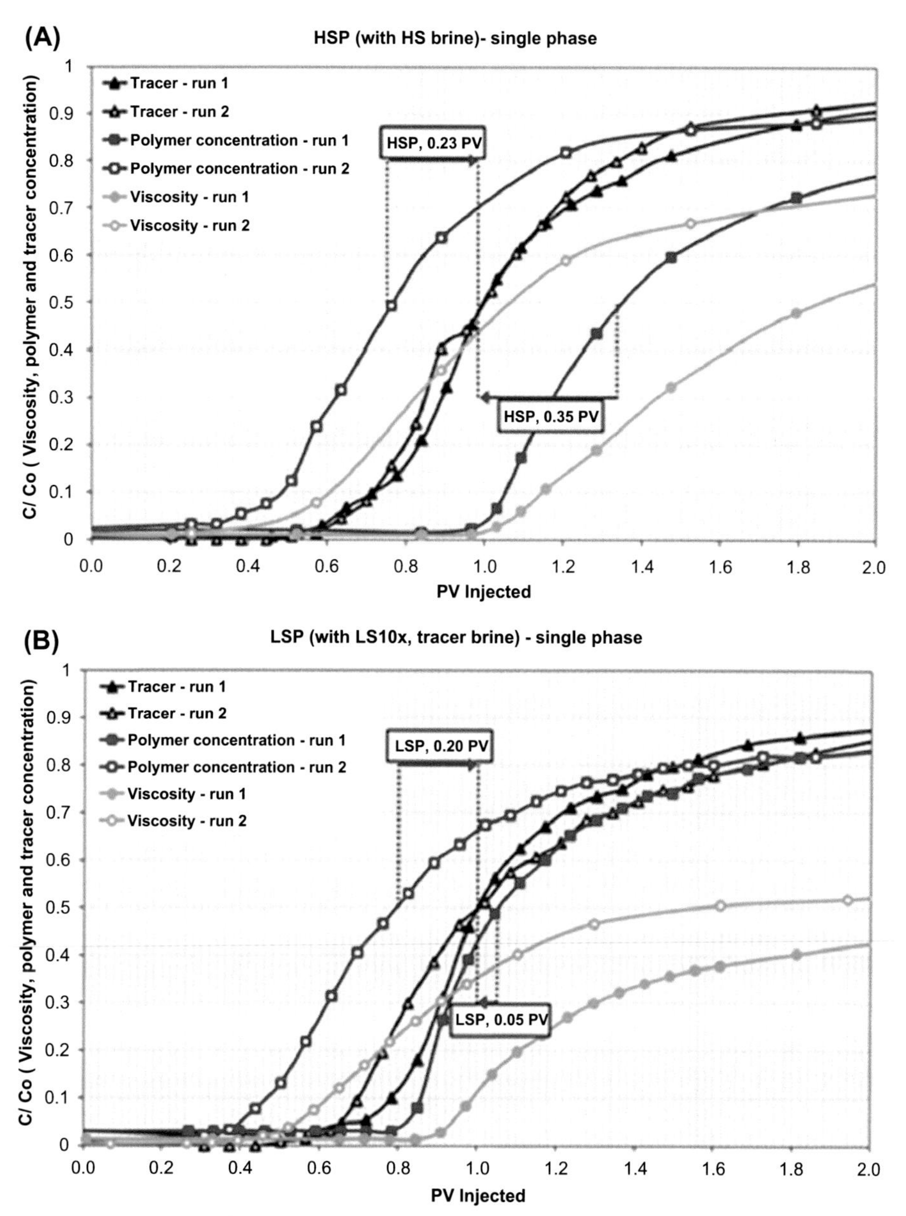

FIG. 4.13 Description of effluent normalized viscosity, and concentrations of polymer and tracer for coreflooding tests of **(A)** conventional polymer flood and **(B)** low-salinity polymer flood. (Credit: From Unsal, E., ten Berge, A. B. G. M., & Wever, D. A. Z. (2018). Low salinity polymer flooding: Lower polymer retention and improved injectivity. *Journal of Petroleum Science and Engineering, 163*, 671–682. https://doi.org/10.1016/j.petrol.2017.10.069.)

injections in the sequence of decreasing salinity follow. Lastly, the formation water is reinjected. In the three sets of coreflooding, the first cycle injecting formation water meets the rupture pressures of 5.03, 6.44, and 3.10 kPa/cm, respectively. After the rupture pressure, the differential pressure halves across the core and becomes stabilized. The initiation of second cycle injecting low-salinity water with 1000 ppm abruptly increases the differential pressure. The third cycle injecting low-salinity water with 5000 ppm shows the higher differential

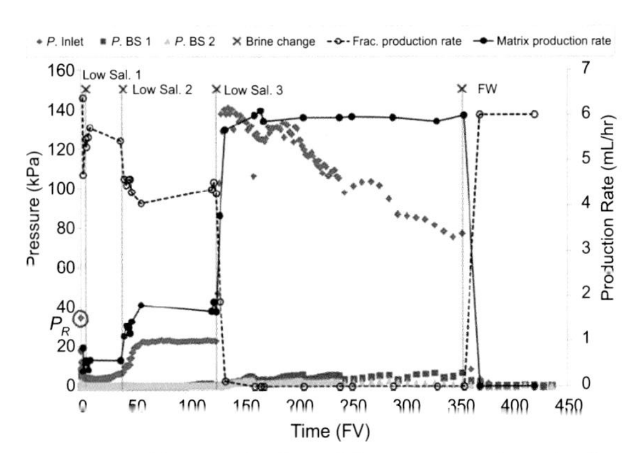

FIG. 4.14 Measured differential pressure and production rates across the fracture and matrix for the three sets of coreflooding using low-salinity water. (Credit: From Brattekås, B., Graue, A., & Seright, R. (2016). Low-salinity chase waterfloods improve performance of Cr(III)-Acetate hydrolyzed polyacrylamide gel in fractured cores. *SPE Reservoir Evaluation and Engineering*, *19*(02), 331–339. https://doi.org/10.2118/173749-PA.)

pressure increase. The last injection of low-salinity water with 0 ppm provides the highest incremental of differential pressure. The increase of differential pressure indicates the higher gel-blocking capacity, which is attributed to the more gel swelling. As gel swells, the volume of fracture becomes reduced. In the three sets of coreflooding, it is consistently observed that there is a higher increase in the differential pressure as the salinity of injecting brine decreases. The gel swelling improving gel-blocking capacity can be achieved by the decreasing salinity of injecting brine. The last cycle reinjecting formation water investigates whether the effect of low-salinity water on the improvement of gel-blocking capacity is reversible or irreversible. It is obviously observed that the injection pressure becomes to the low level. Less than 10 PV injection of formation water completely eliminates the gel swelling by previous low-salinity water injection. It clearly indicates that the gel swelling by the salinity change is the reversible process. The additional two sets of coreflooding verify the long-term stability of gel-blocking capacity by low-salinity water injection. Because the cores are already water-wet system, the increasing production of residual oil saturation by wettability modification is hardly observed. However, it is clear that low-salinity water is favorable to the gel treatment improving conformance issue.

Brattekås and Seright (2018) further published the hybrid technology of gel treatment with low-salinity waterflood on the recovery of fractured and low-permeable carbonates. The study also was interested in the diversion of flow path and gel-blocking capacity, not wettability modification, when LSWF was deployed as the chasing water. The polymer gel fills the fracture volume reducing the fracture conductivity, but it still allows the fluid flow through the fracture channel. After the gel treatment, LSWF as chasing water injection expects to improve the gel-blocking capacity and reduces the fracture channeling restoring matrix flow. This study tried to demonstrate these expectations by measuring the injection pressure and matrix production rate from coreflooding and visualizing flow paths from positron emission tomography (PET) CT scanner. In the coreflooding, chasing water injection using high-salinity brine follows the gel treatment. The high-saline chasing water undergoes the rupture pressure of gel, and then, negligible production rate from matrix is observed. When the chasing waterflood is switched from high-salinity water to low-salinity water, the significant injection pressure increase is observed, not immediately, but after some induction period. Although the increasing pressure is above the gel rupture pressure by a factor of 2, matrix production rate increases by 35%. This result indicates the diversion of fluid flows from fracture to the matrix. This diversion of fluid flow and improved gel-blocking are also confirmed through the interpretation and visualization by PET-CT scanner. In addition, the study addressed the importance of mineralogy and residual oil saturation on the matrix production rate during the hybrid process.

Alhuraishawy, Abdulmohsin Imqam, Wei, and Bai (2016) investigated the roles of low-salinity water in the wettability modification as well as improving the performance of gel treatment in the fractured carbonate reservoirs. They examined the synergetic performance by combining PPG gel treatment and LSWF in the aspects of the PPG strength, swelling of PPGs, fracture width, wettability, and PPG placing pressure. The swelling ratio of PPG is a high function of water salinity. It is known that the swelling ratio increases with a decrease in salinity. The higher swelling ratio indicates the increasing gel volume as well as the decreasing gel strength. The low level of gel strength makes the particles to flow, easily, through the channels and penetrate into the in-depth of a reservoir.

The preliminary study of spontaneous imbibition test confirms that the LSWF processes using 0.1%, 0.01%, and 0.001% NaCl brines recover the higher oil from oil-wet rock samples than the waterflood using 1% NaCl brine. In another imbibition test using water-wet rock samples, LSWF processes injecting 1% and 0.01% NaCl brines provide the negligible change on the oil recovery. This study proposed the two injection scenarios of coinjection of PPG and LSWF and successive injection of LSWF following PPG. The first injection scenario is designed as pre- and postflush injection using 1% NaCl brine and various PPG injections with 0.01%, 0.1%, and 1% NaCl brines. Firstly, in the displacement tests using oil-wet carbonate rocks, the residual resistance factor of coinjection of LSWF and PPG is evaluated by varying the salinity of gel solvent. The PPG with NaCl brine of 0.1% shows the highest residual resistance factor and that with NaCl brine of 0.01% follows the next. The waterflood after the coinjection scheme shows the higher injecting pressure than the waterflood before the coinjection because of the residual resistance factor by gel blocking. The sensitivity of swelling ratio of PPG by salinity is observed. The low salinity condition relatively shows the higher resistance factor indicating higher gel swelling. However, there is an optimum salinity condition to maximize the residual resistance factor. Next experiment investigates the second scenario of hybrid process. It is designed to apply LSWF following PPG with 1.0% NaCl brine. The PPG with 1.0% NaCl solvent is followed by the 1% NaCl brine as postflush process. The residual resistance factor after the injection is estimated by 9.2. Successively, chasing waters of 0.1% NaCl brine and 0.01% NaCl brine are injected after the postflush. The residual resistance factor after the chasing LSWF using 0.1% NaCl brine increases up to 104, which indicates the low salinity condition reswells the gels. Despite high-saline gel solvent, the injection of low-salinity water significantly increases the residual resistance factor. The second chasing LSWF using 0.01% NaCl brine rises the residual resistance factor up to 130. It is clear that low-salinity water injection as chasing water injection confidently improves the resistance factor of gel treatment regardless of salinity of gel solvent. The higher residual resistance factor of LSWF rises the injection pressures as well as oil recovery.

To quantify the wettability modification effect during the hybrid process of low salinity—augmented gel treatment, same displacement test using the water-wet rock is carried out. The comparison between the tests using oil-wet and water-wet rocks captures a couple of observations. The injection of 1% NaCl brine after PPG injection produces higher residual resistance factor in the water-wet system than in the oil-wet system. However, the chasing water injection of 0.1% NaCl brine results in residual resistance factor of 42 for the water-wet rock and 104 for the oil-wet rock. The chasing LSWF modifies the oil-wet rock toward strongly water-wetness, and it negligibly changes the wetness of water-wet rock, which is in line of preliminary spontaneous imbibition tests. In addition, the placing pressure of PPG and fracture width are investigated. The placing pressure of PPG indicates the maximum pressure to inject PPG for each experiment. Both the PPG placing pressure and fracture width slightly or negligibly affect the oil recovery and residual resistance factor.

Alhuraishawy and Bai (2017) published another experiment of the first injection scenario, coinjection of PPG, and LSWF. The microgel with mesh size of 20—30 is used for the PPG injection in this study. The light crude oil of 36 degrees API and oil-wet Indiana limestone are prepared for the displacement experiments. Firstly, swelling ratio measurements calculate the swelling ratio of the dry microgel varying the solvent with different brines (1%, 0.1%, and 0.01% NaCl). The swelling ratio is defined as the difference between the initial weight of dry microgel and the weight of fully swollen gel divided by the initial weight of dry gel. The increasing swollen gel is visualized as salinity of brine decreases (Fig. 4.15). The various fractured

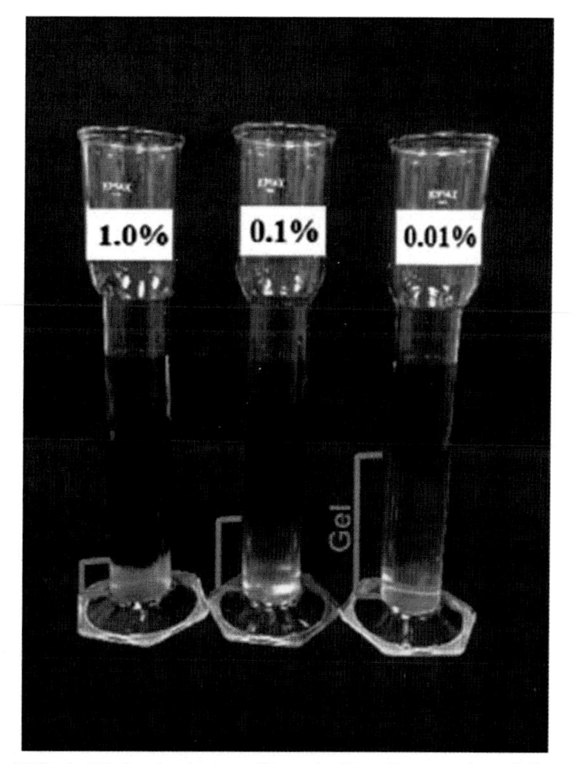

FIG. 4.15 Increasing swollen gel with a decrease in salinity. (Credit: From Alhuraishawy, A. K., & Bai, B. (2017). Evaluation of combined low-salinity water and microgel treatments to improve oil recovery using partial fractured carbonate models. *Journal of Petroleum Science and Engineering, 158*, 80—91. https://doi.org/10.1016/j.petrol.2017.07.016.)

carbonate systems having fracture widths of 0.2, 0.5, and 0.8 cm are subject to the displacement tests. The displacement experiment is designed with pre- and postflushes injecting 1% NaCl brine and PPG injection with various brines of 0.01%, 0.1%, and 1% NaCl. The oil recovery and injection pressure are measured (Fig. 4.16), and the residual resistance factor is calculated from the injection pressure measured.

In the fracture system of fracture width with 0.2 cm, PPG injection with 0.01% NaCl brine recovers higher oil than PPG injections with 0.1% and 1% NaCl brines. This higher incremental oil production is attributed to the higher swelling ratio in lower salinity condition. After the PPG injection with 0.01% NaCl brine, postflush of 1% NaCl brine still increases the oil production. The postflush results in the increasing injection pressure after PPG injection with 0.01% NaCl brine. However, the other PPG injections with 0.1% and 1% NaCl brines lead to the significant drop in the injection pressure for the postflush injection. The PPG injection with higher salinity shows the more decreases in the injection pressure as well as oil recovery for the postflush of 1% NaCl brine. It is clearly confirmed that low salinity condition of the gel solvent is favorable to the gel swelling and residual resistance factor increasing oil recovery and injection pressure. In addition, the swollen gel by low salinity condition is still effective to recover oil and increases the residual resistance factor during the high-

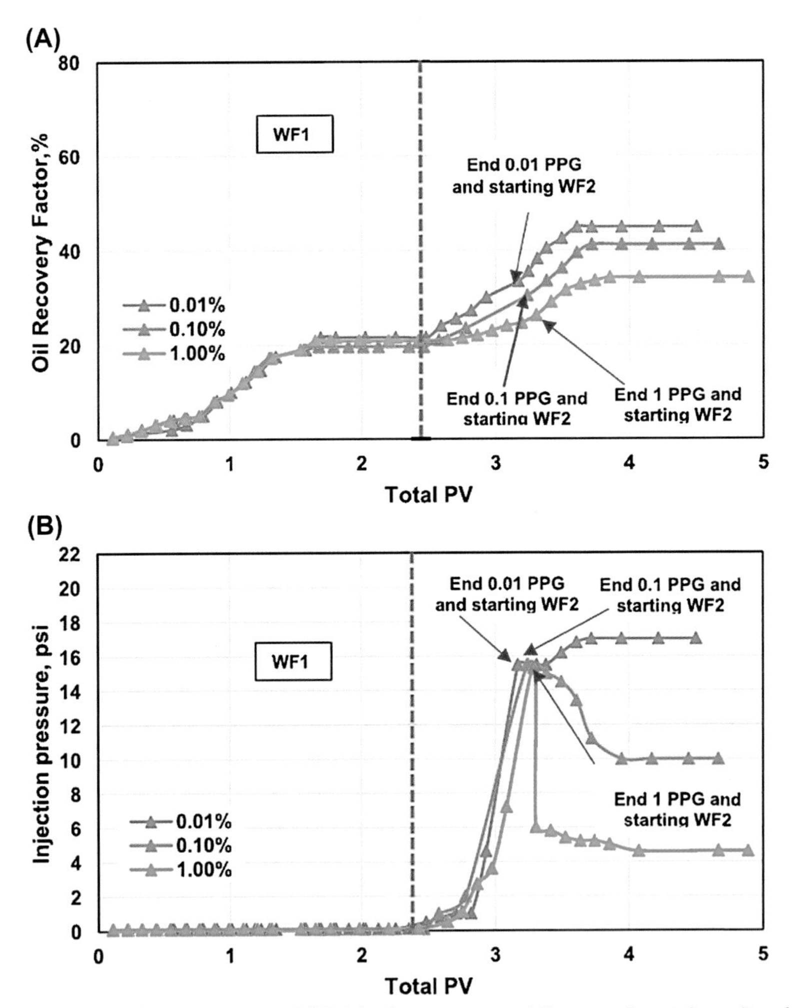

FIG. 4.16 History of **(A)** oil recovery and **(B)** injection pressure of the experimental results of hybrid low salinity–augmented gel treatment. (Credit: From Alhuraishawy, A. K., & Bai, B. (2017). Evaluation of combined low-salinity water and microgel treatments to improve oil recovery using partial fractured carbonate models. *Journal of Petroleum Science and Engineering, 158*, 80–91. https://doi.org/10.1016/j.petrol.2017.07.016.)

saline chasing water injection. In the analysis of fracture width, microgel with 0.01% NaCl brine shows the less incremental oil recovery as chasing water has low salinity. The residual resistance factor shows the equivalent trend. The effect of wettability on the performance of low salinity–based microgel is investigated. The microgel injection with low-salinity water and chasing water is effective to increase oil recovery on both water- and oil-wet reservoirs. Overall, the oil production is higher in water-wet than oil-wet cores. Additional displacement experiment using water-wet core, in which the wettability modification by LSWF is ineffective, confirms the role of low salinity on the gel swelling. The preinjection of 0.01% NaCl and 1% NaCl brines provides negligible difference in the oil recovery, which implies the negligible wettability modification. The PPG and postflush injection using 0.01% NaCl brine show slightly higher recovery of 2% than the injection using 1% NaCl brine. The incremental oil recovery is completely attributed to the higher swelling ratio of microgel in the low-salinity water condition. Lastly, because the higher placing pressure of PPG injects higher volume of microgel, the oil recovery factor and residual resistance factor increase with reducing fracture conductivity. This study demonstrated the effects of low-salinity water injection on the microgel swelling and wettability modification and effects of fracture width and placing pressure of PPG on the performance of hybrid low salinity–based gel treatment.

Alhuraishawy, Bai, and Wei (2018) published the comprehensive experiments of the hybrid process of varying the salinity and potential-determining ion concentration. In the study, the common brines of formation water and seawater are prepared. The formation water has higher salinity than seawater by a factor of about 3. To evaluate the effect of sulfate ion, which is the potential-determining ion in the mechanism of LSWF, and salinity on the performance of the hybrid process, two-times and three-times sulfate-enriched seawaters and the diluted seawaters by factors of 10 and 100 are manufactured. Before the displacement experiment, swelling ratio of microgel is measured for varying brines. All seawater and sulfate-enriched seawaters lead to the gel swelling ratio of 32, and the low-salinity waters of 10-times- and 100-times-diluted seawaters result in the swelling ratios of 120 and 180, respectively. In the displacement experiment, the first test evaluates the effects of low salinity and sulfate ions on the oil recovery from nonfractured carbonate cores. In the test, the two sets of coreflooding are designed with secondary injection of seawater and tertiary injection of low-salinity water or sulfate-enriched seawater, respectively. In the first set of coreflooding, 10-times-diluted and 100-times-diluted seawater injections recover additional 8% and 2% oil, respectively, because of wettability modification effect. The second set also shows that sulfate-enriched seawaters by factors of 2 and 3 produce more oil recovery by 13% and 6.5%, respectively, which are attributed to the wettability modification. The following three displacement tests investigate the oil recovery from fully and partially open-fractured cores. Each displacement test includes a number of coreflooding. The coreflooding is designed as the gel treatment following the preflush of seawater injection, postflush of seawater, and chasing water of low-salinity water or modified seawater injections. In the test with fully fractured core, the first coreflooding evaluates the hybrid gel treatment with LSWF (Fig. 4.17A). The injection of microgel requires significant injection pressure. The postflush of seawater after the gel treatment produces 31% of oil recovery. The chasing water of 10-times-diluted seawater enhances the oil recovery by 13%, and the successive chasing water of 100-times-diluted seawater increases the oil recovery by 26%. Albeit the chasing water using diluted seawater produces some microgel particles, pressure drop is still high. The microgel particle production is the result of the weak gel formation in the low salinity condition. The high injection pressure corresponds to the results of previous swelling ratio measurements, higher swelling ratio of microgel in low salinity condition. Considering the pressure drop, the increasing oil recovery is attributed to both the wettability modification and enhanced gel-blocking capacity by the hybrid process. A couple of corefloodings analyze the performance of the sulfate-enriched seawater injections after gel treatment (Fig. 4.17B). It is consistently drawn that the higher oil recovery is also observed with an increase in the sulfate concentration of seawater. When the gel particle production is not observed, chasing water injection of modified seawater results in the equivalent pressure drop compared with the postflush of seawater. This result indicates the negligible improvement in gel-blocking capacity as well as gel strength during the injection of sulfate-enriched seawater. It is in line with the observation of swelling ratio measurement. Both seawater and sulfate-enriched seawater have same degree of swelling ratio. Other displacement tests also observe similar results. Based on these displacement tests, a couple of conclusions are drawn. The gel-blocking capacity can be secured in sulfate-enriched seawater and enhanced in low-salinity water. In addition, the chasing water injection of low-salinity/modified seawater introduces wettability modification.

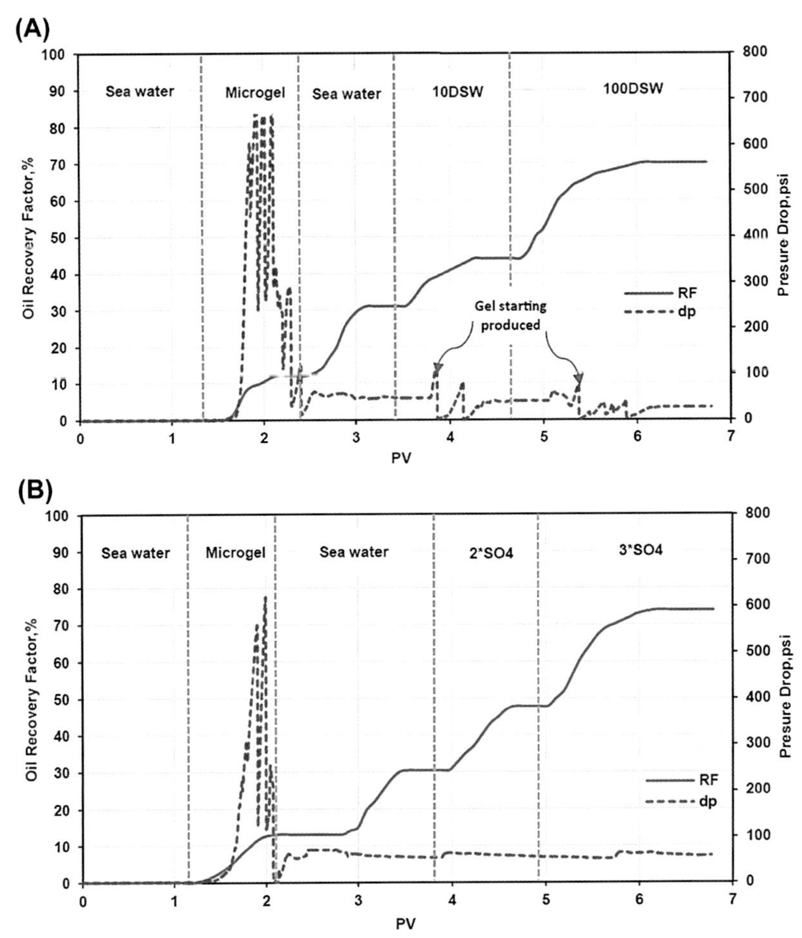

FIG. 4.17 History of oil recovery and pressure drop for hybrid process of **(A)** gel treatment and LSWF, and **(B)** gel treatment and modified seawater injection. (Credit: From Alhuraishawy, A. K., Bai, B., & Mingzhen, W. (2018). Combined ionically modified seawater and microgels to improve oil recovery in fractured carbonate reservoirs. *Journal of Petroleum Science and Engineering*, *162*, 434–445. https://doi.org/10.1016/j.petrol.2017.12.052.)

Therefore, the hybrid process of gel treatment and LSWF or smart waterflood improves the both sweep and displacement efficiencies in fractured carbonate systems.

Numerical simulations: polymer flood

Mohammadi and Jerauld (2012) simulated the hybrid process of polymer flood based on the empirical LSWF modeling. The hybrid LSPF model includes the rheology modeling of polymeric solution, which shows the shear-thinning behavior and chemical degradations by salinity and hardness. In addition, the model considers the adsorption of polymer, permeability reduction, and inaccessible pore volume by the adsorption. The simulation study covers the analysis of oil viscosity, heterogeneity of reservoir, slug size, injection mode, and economics. It is observed that, for more viscous oil reservoir, the oil recovery in the heterogeneous system is close to that in the homogeneous system. This observation implies that the polymer injection of LSPF improves mainly the fractional flow behavior rather than the macroscopic sweep efficiency. In addition, the oil recovery is determined to be higher in the secondary LSPF than the tertiary LSPF. The synergetic behavior of the hybrid process seems to be more effective in tertiary LSPF than secondary LSPF. A synthetic five-spot well simulation using STARS simulator, developed by CMG, confirms the potential of LSPF over the conventional waterflood and polymer flood. Conventional waterflood using high-salinity water produces the ultimate heavy oil recovery of 19.4%. The conventional polymer flood and LSWF increase the oil recovery by 5.6% and 5.9%, respectively. Because the low-salinity water condition remedies the injectivity loss of polymer flood, the injectivity constraint of the process limits the oil recovery by less injection. As a result, the LSPF recovers 10% additional oil over conventional waterflood because of the synergy of wettability modification and mobility ratio improvement. The synergy can be enhanced if LSPF is not involved with injectivity constraint consideration. Additional simulations investigate the potential of infill drilling on the performance of hybrid LSPF and observe the more oil production by the infill drilling.

Khorsandi, Qiao, and Johns (2017) reported the analytical solution of LSPF considering the cation-exchange reaction, wettability modification, adsorption, inaccessible pore volume, and salinity-dependent behavior of polymeric solution for sandstone reservoirs. The study simulated the LSPF model using the in-house compositional simulator, PennSim. The modeling of LSPF assumes that the wettability modification underlying LSWF mechanism is caused by cation-exchange reaction. In detail, the adhered amount of Na^+ on the clay surface controls the relative permeability and capillary pressure of Brooks-Corey model and residual oil saturation. Although the rheology model of polymeric solution takes the polymer concentration and salinity into consideration, it neglects the mechanical degradation by shear rate and viscoelastic behavior of polymeric solution. The polymer rheology model incorporates the residual oil reduction by capillary number or trapping number. The study developed the analytical solution for LSPF as well as LSWF and tried to match a number of experimental results (Seccombe, Lager, Webb, Jerauld, & Fueg, 2008; Shaker Shiran & Skauge, 2013) using the simplified analytical solution. For the LSPF experiment of Shaker Shiran and Skauge (2013), a number of properties including CEC, residual oil saturation at high salinity threshold condition, and residual oil saturation reduction by polymer using the simplified analytical solution are tuned to match the experimental result of oil recovery. Another LSWF simulation attempts to match the LSWF experiment of Seccombe et al. (2008). The simulation refers the data of relative permeability provided from the experiments, and only CEC is tuned as the history-matching parameter. The simulation result accurately reproduces the no oil recovery for the small slug injection because of the mixing. This study successfully developed the analytical solutions incorporating the geochemistry-induced wettability modification and comprehensive rheology model of polymeric solution. The additional trial of LSPF will be described in the numerical simulation of alkali/surfactant/polymer flood.

SURFACTANT FLOOD

Backgrounds of Surfactant Flood

Surfactant EOR process is described to understand the experimental and numerical studies of low salinity–augmented surfactant flood. The backgrounds are summarized from a couple of references (Lake, 1989; Sheng, 2011). The EOR process of surfactant flood injects surface-active agents or surfactants, which are organic compounds to reduce the IFT between the liquid and surfactant and residual oil saturation. The surfactant is composed of tail, which is a nonpolar and hydrophobic hydrocarbon chain, and head, which is a polar hydrophilic group. The amphiphilic surfactant is soluble in both organic solvents and water. The balance between the hydrophilic of head group and hydrophobic of tail part determines the characteristics and type of surfactants. The hydrophilic head group interacts with water, and the hydrophobic tail part interacts with organic solvent, i.e., oil. These interactions form the water-in-oil and oil-in-water microemulsions. When the surfactant is adsorbed at a surface of solid or concentrated at an interface between fluids, the interfacial and surface energies significantly decrease, i.e., IFT/surface tension reductions. There are primary surfactant and cosurfactant to distinguish their roles in surfactant EOR process. The primary surfactant directly forms the microemulsion, and the cosurfactant augments the activities of the primary surfactant modifying the surface energy, the viscosity of the liquids, etc. The surfactants are, conventionally, classified as anionic, cationic, nonionic, and zwitterionic surfactants based on the ionic nature of the hydrophilic head group. Because the anionic surfactant exhibits the relatively low adsorption on negatively charged clay, it is widely used for sandstone reservoirs. Although the nonionic surfactant has higher tolerance to the salinity, its ability reducing IFT is not sufficient as anionic surfactant. Cationic surfactant can be used for the carbonate reservoirs rather than the sandstone reservoir because of higher adsorption in sandstone reservoir. Because the zwitterionic surfactants have two active groups, they are classified as nonionic/anionic, nonionic/cationic, and anionic/cationic surfactants. The expensive zwitterionic surfactants have higher tolerance to the temperature and salinity. Most of surfactants used in the application of surfactant EOR process are the sulfonated hydrocarbons such as alcohol propoxylate sulfate and alcohol propoxylate sulfonate.

Hydrophile-lipophile balance

There are a number of methods to characterize surfactants. One of the methods is the estimation of hydrophile-lipophile balance (HLB), which describes the tendency to solubilize in oil or water. The HLB value indicates the tendency of surfactant to form water-in-oil or oil-in-water emulsion. The HLB is defined using a couple of approaches. The studies (Graiffin, 1954; Griffin, 1949) have defined the HLB value using the molecular mass of hydrophilic relative to the total molecular mass as shown in Eq. (4.12), and the HLB value varies from 0 to 20.

$$\text{HLB} = \frac{20\text{MW}_h}{\text{MW}} \tag{4.12}$$

where HLB indicates the value of HLB; MW_h is the molecular mass of the hydrophilic portion of the molecule; and MW is the total molecular mass of the molecule.

The zero of HLB value indicates the completely hydrophobic molecule of surfactant, and the value of 20 corresponds to the completely hydrophilic/lipophilic molecule of surfactant. The other approach proposed by Davies (1957) considers the degree of the effect of hydrophilic group and defines the HLB value with Eq. (4.13).

$$\text{HLB} = 7 + n_h H_h - n_l H_l \tag{4.13}$$

where n_h is the number of hydrophilic groups; H_h is the value of the hydrophilic groups; n_l is the number of lipophilic groups; and H_l is the value of lipophilic groups.

Critical micelle concentration, Krafft temperature, and cloud point

Another important characteristic is the critical micelle concentration (CMC), which is defined as the concentration of surfactants when the micelles spontaneously form. The addition of surfactant reduces the interface energy and removes the hydrophobic groups of the surfactant from contact with water (Fig. 4.18A). As a result, the free energy of the system decreases. When the surfactant concentration subsequently increases in the system, the surfactant molecules start to aggregate with other molecule and micelles form (Fig. 4.18B). The formation of micelles sharply reduces the free energy of the system by decreasing the contact between the hydrophobic groups of the surfactant and water, until the surfactant concentration reaches the CMC. The increasing concentration of surfactant above the CMC results in more formation of micelles but hardly reduces the free energy of the system. Another factor to be related to CMC is Krafft temperature or critical micelle temperature, which is defined as a minimum temperature where surfactants can form micelles. Below the Krafft temperature condition, no micelles form regardless of the concentration of surfactant. The cloud point is another parameter to illustrate the behavior of nonionic surfactant. It is defined as the temperature, at which the phase

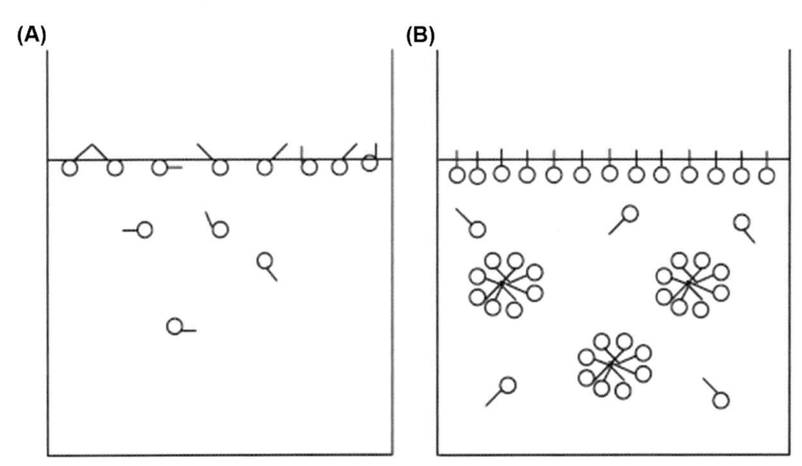

FIG. 4.18 Description of distribution of surfactant molecules **(A)** below and **(B)** above CMC. (Caption: From Sheng, J. (2011). *Modern chemical enhanced oil recovery: Theory and practice*. Amsterdam, Boston, MA: Gulf Professional Pub.)

separation is observed and the solution becomes cloudy. The nonionic surfactants have polyoxyethylene chains, which exhibit a reverse solubility in water depending on the temperature. Because the hydrophilic group of nonionic surfactants has an oxygen, the solubility into water is mainly attributed to hydrogen-oxygen bond. Because the high temperature increases the surfactant molecular activity, the bond becomes weak. Therefore, the high temperature condition makes the cloudy solution because of the separated surfactant molecules.

Phase behavior of microemulsion

It is necessary to distinguish the microemulsion from the macroemulsion beforehand. The macroemulsion is the mixture of two or more immiscible liquids. The one liquid phase is dispersed in the other phase. The dispersed phase in the other continuous phase is commonly appeared to be cloudy. In contrast, the microemulsion system has clear transparency because of the small size of the dispersed emulsion. The aggregation of micelles in the microemulsion system often shows some cloudy transparency. It is thermodynamically stable; therefore, the simple mixing of two or more immiscible liquids forms the microemulsion. The microemulsion is of interest to the surfactant EOR process. The two immiscible phase systems, i.e., oleic and aqueous phases, can be solubilized by the micelles, and the system is termed as the microemulsion system. The microemulsion system is beneficial to the EOR process by recovering trapped oil.

The phase behavior of microemulsion is sensitive to a number of factors including the surfactant type, surfactant concentration, cosolvent, oil composition, presence of alkali, salinity, temperature, and to much lesser degree, pressure. Because no universal equation exists to model the phase behavior of microemulsion considering the factors, the phase behavior has to be identified with experiments. Among the factors, the salinity is of importance because of the huge influence to change phase behavior of the system. Conventionally, the solubility of the anionic surfactant in the brine decreases with an increase in the salinity of brine, i.e., the surfactant is separated from the aqueous phase as the electrolyte concentration increases. In turn, the surfactant easily dissolves in oleic phase rather than in aqueous phase at high salinity condition. The microemulsion system of high salinity condition is the oil-external microemulsion and excess water system. Because the microemulsion in oleic phase has less density than water, the microemulsion is located above the water. At the low salinity condition, the surfactant shows the higher solubility in aqueous phase and negligible solubility in oleic phase. The microemulsion system of low salinity condition is the water-external microemulsion and excess oil system. The microemulsion in aqueous phase is denser than excess oil and it resides below the oleic phase. In the intermediate salinity condition, the microemulsion system is involved with three phases of excess oleic phase, microemulsion, and excess aqueous phase. In this system, the density of microemulsion is between the densities of oleic and aqueous phases. The microemulsion is located above the aqueous phase and below the oleic phase. The phase behavior of the microemulsion system at different salinity conditions is described in Fig. 4.19. There are a couple of terminologies to indicate the microemulsion type. The Winsor type I or type II (−) indicates the water-external microemulsion in low salinity condition. The Winsor type II or type II (+) indicates the oil-external microemulsion in high salinity condition. In the intermediate salinity condition, Winsor type III or type III forms.

There are terminologies to determine the phase behavior of the microemulsion and type of microemulsion: (1) solubilization ratio, (2) R-ratio, and (3) packing factor. The solubilization ratio is defined as the ratio of the solubilized volume of oil or water to the surfactant volume in the microemulsion system as represented in Eqs. (4.14) and (4.15). Healy, Reed, and Stenmark (1976) developed the relationship between solubilization ratios and interfacial tension between the microemulsion and excess phase. It is also suggested that the equal amounts of oil and water are solubilized in the microemulsion Winsor type III, indicating the optimum solubilization ratio, at optimal salinity. Huh (1979) theoretically developed the correlation between the interfacial tension and optimum solubilization ratio as shown in Eq. (4.16). When the optimum solubilization ratio is higher than 10, the IFT at the optimum salinity is on the order of 10^{-3} dyne/cm or less and it sufficiently mobilizes the residual oil saturation.

$$S_o = \frac{V_o}{V_{surf}} \tag{4.14}$$

$$S_w = \frac{V_w}{V_{surf}} \tag{4.15}$$

$$\sigma = \frac{C}{S^2} \tag{4.16}$$

where S_o is the oil solubilization ratio; V_o is the solubilized volume of oil in microemulsion; V_{surf} is the surfactant volume in microemulsion; S_w is the water solubilization ratio; V_w is the solubilized volume of water in microemulsion; C is the constant and,

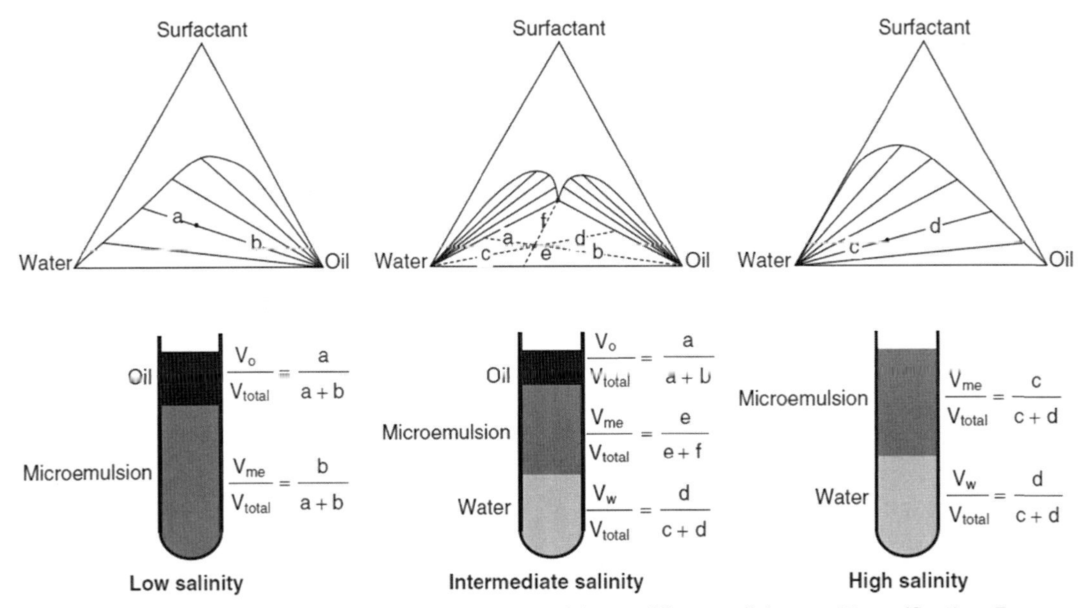

FIG. 4.19 Description of phase behavior of microemulsion at different salinity conditions. (Caption: From Sheng, J. (2011). *Modern chemical enhanced oil recovery: Theory and practice*. Amsterdam, Boston, MA: Gulf Professional Pub.)

approximately, 0.3 dyne/cm for typical crude oils and surfactant; and S is the optimum solubilization ratio.

Winsor (1948) proposed the R-ratio considering the affinity of surfactant to aqueous and oleic phases in the interfacial zone. The interfacial zone has a limited thickness separating the bulk of aqueous phase from the bulks of oleic phase. The zone includes hydrophilic heads, lipophilic tails, and oil and water molecules. In the interfacial zone, there are a number of cohesive energies among surfactant, and water and oil molecules. The cohesive energy between surfactant and oil molecules is attributed to the interactions of oil molecules with both hydrophilic heads and lipophilic tails. The cohesive energy between surfactant and water molecules consists of the interactions of water molecules with both hydrophilic heads and lipophilic tails. The R-ratio is defined as the ratio of the energy of surfactant-oil molecules to the energy of surfactant-water molecules as shown in Eq. (4.17). Considering the repulsive interactions in oil molecules, water molecules, and surfactant molecules, respectively, Bourrel and Schechter (1988) proposed the extended version of the R-ratio as Eq. (4.18).

$$R = \frac{A_{co}}{A_{cw}} \tag{4.17}$$

$$R = \frac{A_{co} - A_{oo} - A_{ll}}{A_{cw} - A_{ww} - A_{hh}} \tag{4.18}$$

where R is the R-ratio; A_{co} is the cohesive energy of surfactant-oil molecules; A_{cw} is the cohesive energy of surfactant-water molecules; A_{oo} is the energy of repulsive interaction in oil molecules; A_{ww} is the energy of repulsive interaction in water molecules; A_{ll} is the energy of repulsive interaction in the lipophilic groups of surfactant molecules; and A_{hh} is the energy of repulsive interaction in the hydrophilic groups of surfactant molecules.

When the R-ratio is less than one, the relative miscibility with water increases and that with oil decreases, indicating the Winsor type I. Winsor type II shows higher R-ratio than one. When the ratio is equal to one, the system is in Winsor type III of microemulsion system. Another terminology of packing factor also describes the microemulsion type.

The microemulsion is sensitive to temperature and salinity and negligible to pressure. Because the injecting brine and formation water have different salinities, mixing between injecting brine and formation water easily occurs in the porous media. Considering the mixing process, the phase behavior test, i.e., pipette test, identifies a salinity-dependent volume fraction diagram or salinity-dependent solubilization ratio at the specific temperature to determine optimum salinity of microemulsion system. The salinity-dependent volume fraction diagram calculates the salinity-dependent solubilization ratios of water and oil using Eqs. (4.14)

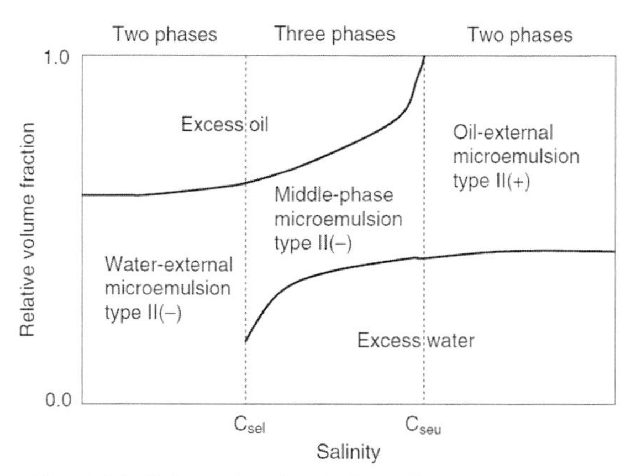

FIG. 4.20 Schematic description of salinity-dependent volume fraction as salinity changes. (Caption: From Sheng, J. (2011). *Modern chemical enhanced oil recovery: Theory and practice*. Amsterdam, Boston, MA: Gulf Professional Pub.)

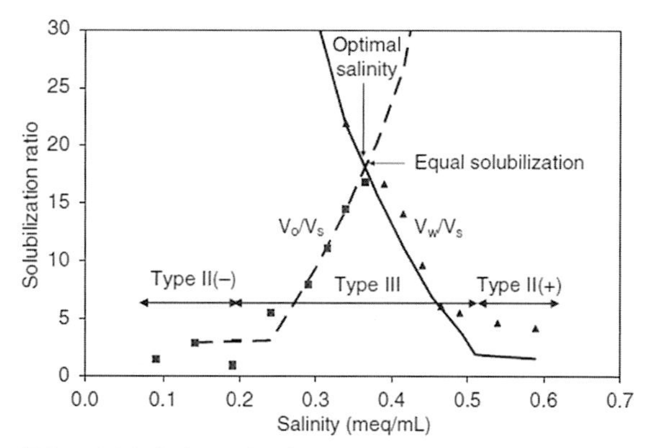

FIG. 4.21 Schematic description of salinity-dependent solubilization ratio of oil and water as salinity changes. (From Sheng, J. (2011). *Modern chemical enhanced oil recovery: Theory and practice*. Amsterdam, Boston, MA: Gulf Professional Pub.)

and (4.15). The salinity-dependent solubilization ratio easily determines optimum salinity of the microemulsion system. In the microemulsion system, the optimum salinity is identified at which both water/microemulsion and oil/microemulsion incorporate the minimum IFT values. The solubilization ratio of water becomes equal to the ratio of oil at the optimum salinity. It is described that the microemulsion changes from Winsor type I to Winsor type III and from Winsor type III to Winsor type II as the salinity, i.e., electrolyte, increases. Fig. 4.20 schematically depicts the relative volume fraction diagram of respective Winsor type of microemulsion as salinity changes. The relative volume of respective Winsor type of microemulsion is experimentally measured. Fig. 4.21 shows the solubilization ratio of oil and water based on the relative volume fraction diagram. Therefore, the phase behavior test suggests the optimum formula of injecting salinity, considering mixing the injecting brine and formation water, to be close to optimum salinity condition of the in situ microemulsion system.

The phase behavior of microemulsion is complex and sensitive to various factors. The ternary diagram (Fig. 4.22) and empirical correlation of Hand's rule are useful to represent and interpret the phase behavior. Because the microemulsion consists of a number of components, including water and electrolytes, hydrocarbon, surfactant, and cosurfactant or alcohol, the terminology of pseudocomponent is introduced. The ternary diagram is formulated with the three pseudocomponents of water, oil, and surfactant. The ternary diagram illustrates the phase and overall compositions as well as relative amounts using a number of terminologies: tie line, binodal curve, connodals, invariant point, etc. The tie line connects the compositions of two equilibrium phases at its two ends. The binodal curve separates the one-and two-phase region. The one-phase region is above the binodal curve, and the two-phase region is below the curve. The empirical correlation of Hand's rule represents relationship between the tie lines and binodal curves (Hand, 1929).

Retention

The surfactant flood has a risk of expensive surfactant loss by retention. Identification of the retention is one of the crucial assessments for the successful surfactant EOR process. The retention of surfactant includes the precipitation, adsorption, and phase trapping. Because the solubility of surfactant is sensitive to the salinity, the surfactant can precipitate or aggregate depending on the salinity. In addition, the cosolvent alcohol can increase the solubility of surfactant to prevent precipitation or aggregation problems. The adsorption of surfactant on the rock surface loses the surfactant concentration from the bulk solution. The adsorption on the rock surface varies by the type, equivalent weight, and surfactant concentration of surfactant, rock mineral, clay content, salinity, temperature, pH, flow rate, etc. The retention by phase trapping is attributed to the mechanical trapping, phase partitioning, or hydrodynamical trapping.

Mechanisms of surfactant EOR process

The main mechanism of surfactant EOR process is to reduce IFT. The IFT reduction is highly related to the

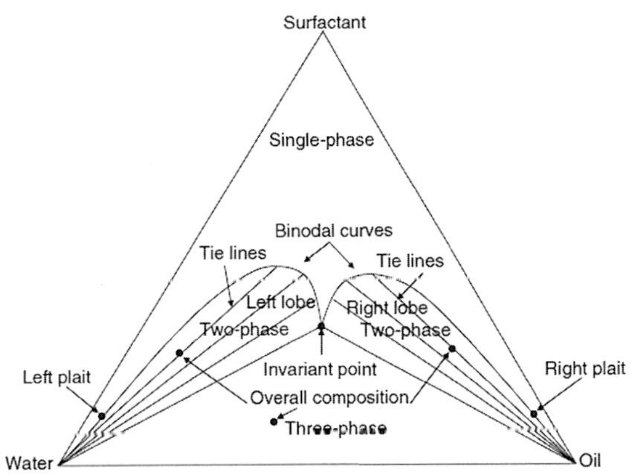

FIG. 4.22 Schematic description of ternary diagram. (From Sheng, J. (2011). *Modern chemical enhanced oil recovery: Theory and practice*. Amsterdam, Boston, MA: Gulf Professional Pub.)

phase behavior of microemulsion and solubilization ratios of water and oil. A couple of explanations are proposed for the IFT reduction. The adsorption of surfactant molecules on the oil/water interface and the formation of micelles reaching CMC potentially achieve the ultralow IFT overcoming the capillary forces holding the oil in reservoirs. The concept of capillary number is discussed to explain the relationship between IFT reduction and oil recovery increase. The capillary number is the ratio of viscous force to capillary force. The reduction in IFT decreasing capillary force increases the capillary number. The increasing capillary number implies the relatively higher viscous force over capillary force and leads to the residual oil to be mobilized. Another important mechanism of surfactant flood is to improve the initial wettability of reservoirs when the surfactant interacts with the rock surface. Incorporating the mechanisms, the surfactant EOR process, conventionally, can be categorized into diluted surfactant flood and micellar flood. Because of the expensive cost of surfactant additives, it is practically the immiscible process in the field application. Fully miscible process might be achieved in the early stage of the process, but the immiscible process quickly follows forming the multiple phases.

Optimum salinity gradient

During the transport of surfactant solution in porous media, the retention of surfactant decreases the surfactant concentration in bulk solution. In addition, the surfactant concentration decreases as the surfactant solution moves forward in porous media. The change in the surfactant concentration varies the in situ optimum salinity condition of microemulsion. If the phase behavior test determines the surfactant to have tendency of decreasing optimum salinity as the surfactant concentration decreases, the injecting surfactant solution with decreasing salinity might maintain the optimum condition during the transport in porous media. It is termed as negative salinity gradient, which indicates the salinities of preflush water, surfactant slug, and postflush are in descending order. Nelson (1982) published experimental works to demonstrate the negative salinity gradient achieving the optimum salinity condition. Whether optimum salinity gradient is positive or negative depends on the system considering the surfactant, cosolvent, salinity, divalent cations, etc. Therefore, it is clear that the achievement of ultralow IFT at optimal salinity increases the oil production but does not guarantee the highest oil recovery in real field applications. The comprehensive investigations of surfactant for a specific condition are necessary for successful application of surfactant EOR.

Low Salinity–Augmented Surfactant Flood

Experiment

Alagic and Skauge (2010) proposed the hybrid EOR process when the LSWF is combined with surfactant flood. The study investigated the tertiary low salinity–augmented surfactant flood (LSSF) following secondary LSWF or secondary conventional waterflood. In addition, it quantified the role of high pH condition and alkaline effect on the performance of tertiary LSSF. The low salinity condition expects to bring a couple of advantages including the improving solubility of surfactant and reduction of retention to surfactant EOR process. In the experiments, the synthetic seawater of 36,321 ppm TDS and low-salinity water of 0.5% NaCl brine are investigated. The anionic surfactant is prepared and will form the Winsor type 1 microemulsion with the low-salinity water. The study carried out the IFT measurement, coreflooding, ion analysis using inductively coupled plasma (ICP), etc. The viscosities of the surfactant solutions/brines and the IFTs between crude oil and surfactant solutions/brines are measured. The viscosity of low-saline surfactant solution is slightly higher than surfactant-free brines. The IFT between crude oil and seawater shows the 23.5 dyne/cm. Moderate IFT reduction by 9 dyne/cm is observed for the low-salinity water. When the pH of low-salinity water increases by the addition of NaOH, the IFT decreases to 1.8 dyne/cm. The low-saline surfactant solution achieves the low IFT on the order of 10^{-2} dyne/cm at moderate and high pH conditions. The higher pH shows the lower IFT for the low-saline surfactant solution, but the change is not significant. Using the fluids, the four sets of coreflooding are performed. Two coreflooding experiments are designed with secondary LSWF and tertiary LSSF. Additional oil recoveries by about 32% and 30% are obtained by tertiary LSSF process, respectively. The increasing differential pressure and pH are observed during the tertiary LSSF. In the third coreflooding, the secondary injection of seawater is followed by the tertiary mode of LSSF. The tertiary LSSF enhances the oil recovery with the increasing pH and differential pressure. These observations are explained that the increasing pH is attributed to the alkaline properties of the surfactant solution. The incremental oil recovery is higher in previous two coreflooding experiments compared with the third coreflooding experiment. The last coreflooding investigates the performance of secondary LSWF and tertiary high-pH LSWF. The slight increment of oil recovery by 7% is achieved for the tertiary process. The experimental study drew a number of conclusions. The IFT reduction of surfactant is still effective in low salinity condition. Stabilizing a low salinity environment can improve the tertiary oil recovery of LSSF. The contribution of pH increase by alkali addition is slightly effective for tertiary recovery of LSWF.

Alagic, Spildo, Arne, and Jonas (2011) further investigated the performance of tertiary LSSF on the recovery of remaining oil after secondary LSWF and quantified the role of crude oil aging and initial wettability on the oil recovery of LSWF and LSSF. The same anionic surfactant and two different concentrations of surfactant are prepared. The coreflooding experiments use the aged cores, i.e., less water-wet, and unaged cores, i.e., water-wet. During the coreflooding of secondary LSWF, the retardation of Mg^{2+} is observed for the aged core. Because the mineral dissolution might compensate the retardation of Ca^{2+}, the retardation of Ca^{2+} is hardly measured for both aged and unaged cores. The less water-wet core, aged core, shows the less fine migration than the water-wet core, unaged core. Less fine migration is consistent with extensive studies of LSWF. The secondary LSWF also produces higher oil recovery in aged cores than unaged cores. During the coreflooding of tertiary LSSF, the higher recovery of remaining oil is obtained for high surfactant concentration and crude oil–aged core system. During hybrid LSSF, the oil layer becomes destabilized at low salinity condition and lower capillary pressure by surfactant injection mobilizes the trapped oil. The studies (Alagic et al., 2011; Alagic & Skauge, 2010) experimentally have demonstrated the potential of hybrid LSSF on recovery from the less water-wet cores.

Johannessen and Spildo (2013) investigated the potential of LSSF by comparing the optimum salinity surfactant flood. The LSSF introduces the moderately low IFT, and the surfactant flood at optimum salinity brings the ultralow IFT. Two North Sea crude oils and the various versions of diluted seawater are prepared for the experiments. The experiments include the phase behavior test of microemulsion, IFT measurement, coreflooding, dispersion test, and dynamic retention measurement. The less water-wet Berea core by crude oils aging is used for coreflooding. The low-salinity brine is the diluted seawater by a factor of 0.07. The microemulsion system is composed of the water, surfactant, cosurfactant, cosolvent, and crude oil. The solubilization ratio and IFT by varying salinity and WOR are experimentally measured at 50°C through phase behavior test. The optimum salinity corresponding one of the crude oils is determined as diluted seawater by a factor of 0.43. The optimum salinity of surfactant solution is relatively higher than the low salinity condition, which is diluted seawater by a factor of 0.7. The surfactant solution at optimal salinity condition is also observed to have ultralow IFT on the order of 3×10^{-4} dyne/cm. The surfactant solution at low salinity condition results in the moderately low IFT as 1.8×10^{-2}. Another crude oil is determined to have the optimum salinity condition as diluted seawater by a factor of 0.5. In a number of coreflooding tests, LSSF and optimum salinity surfactant flood process are compared (Fig. 4.23). The hybrid LSSF is designed to follow the secondary or tertiary LSWF. The optimized surfactant flood also follows the secondary or tertiary optimum salinity waterflood. Significant increases in oil recovery are observed when surfactant is injected regardless of salinity condition. The additional oil recoveries by LSSF range from 7% to 30%. The increments by 23.2%–37% are observed during optimum salinity surfactant flood. The relatively lower oil recovery increase of LSSF than optimum surfactant flood is attributed to the higher performance of preflush LSWF compared with that of preflush optimum salinity waterflood. The dispersion test is carried out to examine the heterogeneity of cores. The core cleaning process differentiates the dispersion profiles from cores. It is explained that residual oil is blocking the pore and cleaning process eliminating the residual oil gives accessibility to the isolated and dead-end pore. Lastly, the retention of surfactant is measured to validate the benefit of low salinity condition. The dynamic measurement of retention monitors the produced surfactant concentrations of coreflooding experiments. The less retardation and higher total production of surfactant

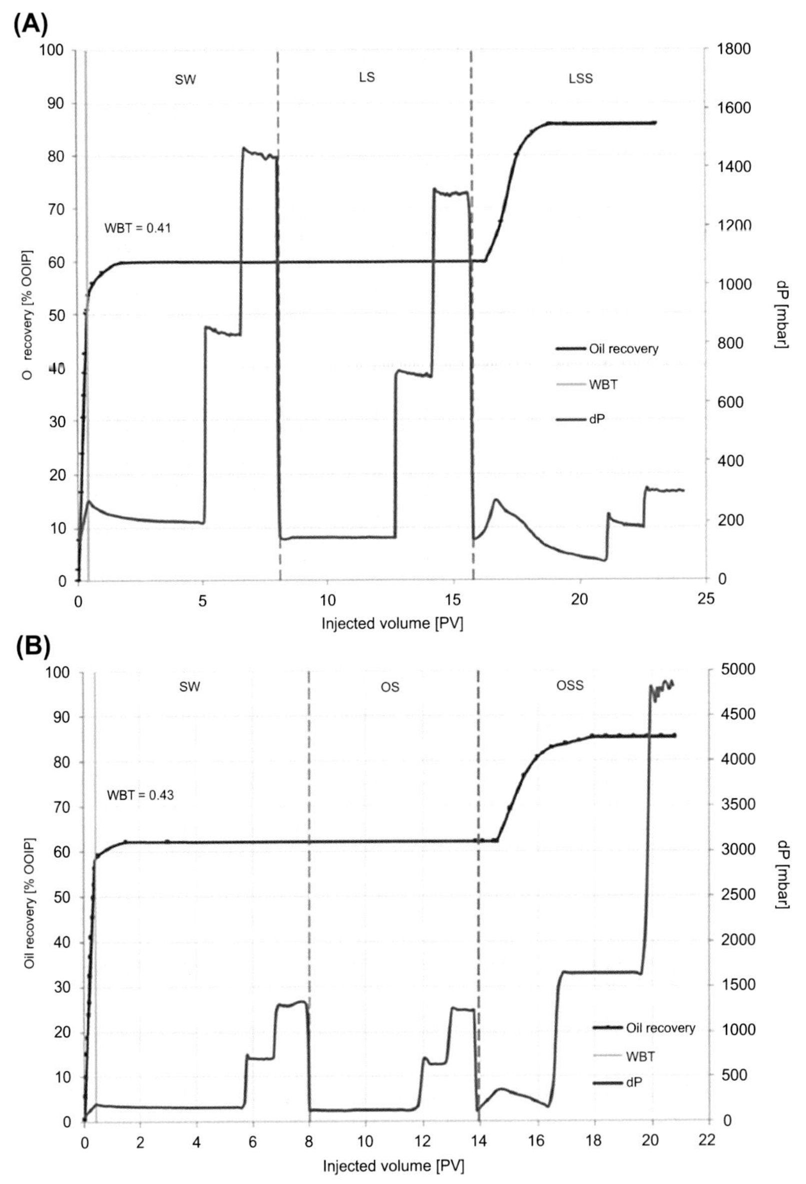

FIG. 4.23 The comparison between corefooding experiments of **(A)** hybrid low salinity surfactant flood and **(B)** optimum salinity surfactant flood. (Credit: From Johannessen, A. M., & Spildo, K. (2013). Enhanced oil recovery (EOR) by combining surfactant with low salinity injection. *Energy and Fuels*, *27*(10):5738–5749. https://doi.org/10.1021/ef400596b.)

are observed during LSSF compared with the optimum salinity surfactant flood (Fig. 4.24). The average retentions are 0.39 mg surfactant/g rock for optimum salinity surfactant flood and 0.24 mg surfactant/g rock for LSSF. It is concluded that the hybrid LSSF produces higher oil recovery comparable with the optimum salinity surfactant flood with less surfactant retention.

Khanamiri, Torsæter, and Stensen (2015) investigated whether the hybrid process of LSSF or the combination process of LSWF and optimum salinity surfactant flood injection shows higher EOR potential. The in situ brine, NaCl brine, and KI brine are manufactured for Berea sandstone coreflooding. The surfactant of sodium dodecylbenzenesulfonate (SDBS) is used to

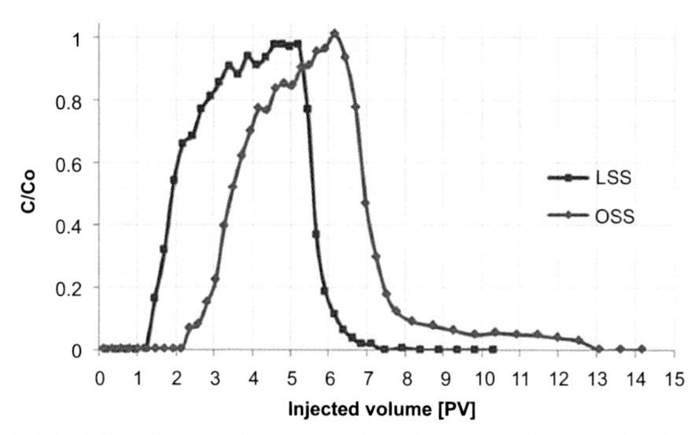

FIG. 4.24 Comparison of produced surfactant concentration between hybrid low salinity surfactant flood and optimum salinity surfactant flood. (Credit: From Johannessen, A. M., & Spildo, K. (2013). Enhanced oil recovery (EOR) by combining surfactant with low salinity injection. *Energy and Fuels*, *27*(10): 5738–5749. https://doi.org/10.1021/ef400596b.)

formulate the microemulsion. The experiments include the IFT measurement, solubility test, and coreflooding. Before the coreflooding experiment, the IFT and aqueous solubility of surfactant are examined by changing the surfactant concentration and brine type. The solubility test investigates the precipitation of surfactant. The precipitation might introduce the surfactant loss increasing EOR application cost. The surfactant concentration is increased from 300 to 3000 mg/L. The surfactant solution with 10-times-diluted in situ brine precipitates when the surfactant concentration increases. The IFT measurement shows the lowest IFT of 0.005 dyne/cm when the surfactant solution has the surfactant of 500 mg/L and NaCl brine of 32,500 ppm TDS. These IFT measurement and solubility test indicate that the surfactant solution is the optimum salinity surfactant solution. The makeup brine of low-saline surfactant solution is the 10-times-diluted NaCl brine of 3350 ppm TDS. The low-saline surfactant solution shows no precipitation as well as IFT of 1.6 dyne/cm. The first set of coreflooding experiments is designed with secondary LSWF, tertiary LSSF, and chasing high-salinity waterflood. The second set of coreflooding experiments is designed with secondary high-salinity waterflood or LSWF, tertiary optimum salinity surfactant flood, and chasing LSWF or high-salinity waterflood. The comparison between the two sets showed that the process of secondary LSWF and tertiary LSSF recovers more oil than the process of tertiary optimum-salinity surfactant flood. For the application of optimum-salinity surfactant flood, the design of pre- and postflushes to be either LSWF or high-salinity waterflood is crucial factor for the oil production.

The experimental studies (Hosseinzade Khanamiri, Baltzersen Enge, et al., 2016; Hosseinzade Khanamiri, Nourani, et al., 2016) continued to investigate factors affecting the EOR potential of hybrid LSSF process in Berea sandstone. The studies adjusted the concentration of cations of Na^+, Ca^{2+}, and Mg^{2+} as well as salinity in the composition of brine. The high-salinity in situ brine has 31,061 ppm TDS. The low-salinity water as the 10-times-diluted in situ brine is prepared. There are modified versions of in situ brine and low-salinity water. The modified versions of in situ and low-salinity brines are prepared by reducing the concentration of Ca^{2+} and Mg^{2+}. Each modified brine has the equivalent salinity of initial brine by adding NaCl. Crude oil has the TAN of 1.08 mg KOH/g and TBN of 1.16 ± 0.35 mg KOH/g. The anionic surfactant is used to formulate the microemulsion. The experimental studies carried out the phase behavior test, coreflooding experiment, surfactant adsorption test, and contact angle measurement. Detailed observations of the phase behavior test can be found in the study of Tichelkamp et al. (2016). Controlling the ionic strength of brine and the molar ratio between calcium and sodium ions, the IFTs are measured to demonstrate the EOR potential of low-salinity surfactant solution. The solubility test shows the surfactant precipitation when the low-salinity surfactant solutions have the cations of both Ca^{2+} and Mg^{2+} or only Ca^{2+}, and higher concentration of surfactant above 1000 mg/L. Therefore, the surfactant concentration of 500 mg/L is selected for the coreflooding experiments. In the study of Hosseinzade Khanamiri, Baltzersen Enge, et al., 2016, the coreflooding experiments are designed with the secondary LSWF and tertiary LSSF, changing the makeup brine type. During the tertiary LSSF, the significant increase of oil recovery by tertiary LSSF (Fig. 4.25) and higher endpoint of relative permeability to water are observed (Fig. 4.26). However, the oil recoveries after secondary LSWF and tertiary LSSF vary depending on aging duration and the composition of formation brine and makeup brine. Adsorption experiment using packed bed indicates the less adsorption and less surfactant loss when divalent cation does not exist. Further coreflooding experiment investigates the tertiary mode of optimum-salinity surfactant flood. The capillary numbers of optimum-salinity surfactant flood and LSSF processes are roughly calculated. The capillary number of optimum-salinity surfactant flood shows two or three orders magnitude higher than LSSF. However, the LSSF produces incremental oil recovery comparable with the optimum-salinity surfactant flood.

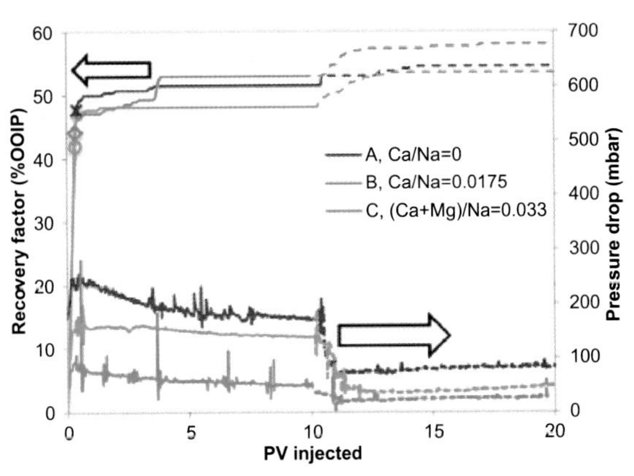

FIG. 4.25 Recovery and pressure drop in the experiments of secondary low-salinity waterflood and tertiary low salinity–augmented surfactant flood. (Credit: From Hosseinzade Khanamiri, H., Baltzersen Enge, I., Nourani, M., Stensen, J.Å., Torsæter, O., & Hadia, N. (2016). EOR by low salinity water and surfactant at low concentration: Impact of injection and in situ brine composition. *Energy and Fuels, 30*(4), 2705–2713. https://doi.org/10.1021/acs.energyfuels.5b02899.)

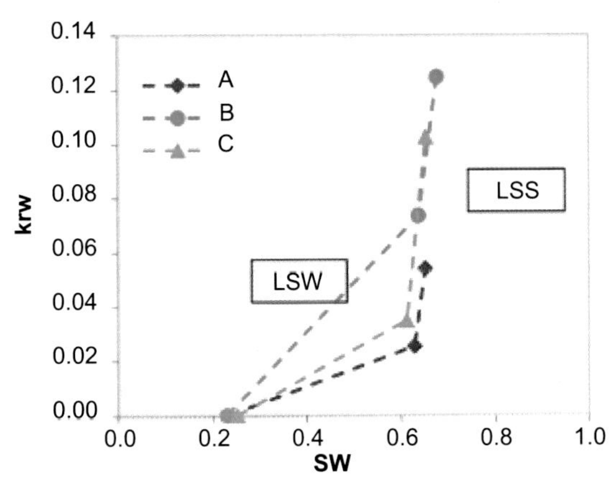

FIG. 4.26 Endpoint relative permeabilities to water in the experiments of secondary low-salinity waterflood and tertiary low salinity–augmented surfactant flood. (Credit: From Hosseinzade Khanamiri, H., Baltzersen Enge, I., Nourani, M., Stensen, J.Å., Torsæter, O., & Hadia, N. (2016). EOR by low salinity water and surfactant at low concentration: Impact of injection and in situ brine composition. *Energy and Fuels, 30*(4), 2705–2713. https://doi.org/10.1021/acs.energyfuels.5b02899.)

Zhang, Yu, Cheng, and Lee (2015) formulated a variety of microemulsions without alkali and investigated the phase behavior and IFT reduction of the microemulsions. Conventionally, the alkali addition decreases the optimal salinity of surfactant solution, but it can introduce risks for chemical EOR process. A number of formulations of microemulsions are prepared using the anionic, cationic, zwitterionic, and nonionic surfactants. Mixtures of different surfactant types have a risk to precipitate because of strong electrical interaction. However, the mixture with appropriate mixing ratio is able to produce benefits without precipitation problem.

Firstly, the surfactant mixture between cationic and ionic surfactants is investigated through phase behavior test of microemulsion and IFT measurement. In the phase behavior test, the mixing ratio between anionic and cationic surfactants varies from 4:1 to 1:4. The cloudy solutions are observed with mixing ratios of 2:1, 1:1, and 1:2. The mixing ratios either higher than 2:1 or smaller than 1:2 show the clear transparency of solution. Using the mixing ratios of 4:1 and 2:1, surfactant solutions are investigated in terms of salinity-dependent volume fraction of microemulsion type. For the surfactant mixture with the mixing ratio of 4:1, anionic surfactant is predominant over cationic surfactant. The phase behavior test evaluates the microemulsion type in the range of 0–24 wt% salinity. The Winsor type III of microemulsion appears in range of high salinity conditions including optimal salinity, 10–24 wt%. The optimal salinity is determined as 19 wt%. The salinity less than 10 wt% shows the Winsor type I. The formulation of Winsor type II is not observed in the range of salinities to be tested. For the surfactant mixture of mixing ratio of 2:1, which has potential to aggregate, Winsor type III is observed in the range of 0–10 wt% salinity. The optimal salinity is determined to be 3 wt%. In contrast to the previous surfactant mixture, Winsor type II is formulated above 10 wt%. No Winsor type I is observed. These phase behavior tests reveal that the mixing ratio controls the salinity condition determining microemulsion type. In addition, the mixing ratio significantly changes optimal salinity (Fig. 4.27). Additional experiments investigate another surfactant mixture using different anionic and same cationic surfactants. The experiments using the two different surfactant mixture solutions indicate that the properties of surfactant mixture are closely related to the selected surfactant type.

Secondly, the formation of surfactant mixture using anionic and zwitterionic surfactants is investigated. Because the zwitterionic surfactant exhibits both positive and negative charges, it behaves like anionic surfactant at high pH and cationic surfactant at low pH. Surfactant mixture is prepared adapting the mixing ratio between anionic and zwitterionic surfactants as 2:1. In the range of 0–18 wt% salinity, only Winsor type III is observed at pH 2 condition. In addition, the volume

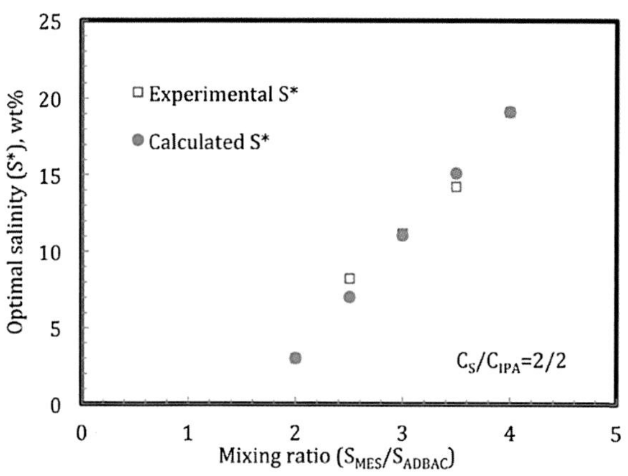

FIG. 4.27 Effect of mixing ratio on optimal salinity. (Credit: From Zhang, G., Yu, J., Cheng, D. U., & Lee, R. (2015). Formulation of surfactants for very low/high salinity surfactant flooding without alkali. *Paper presented at the SPE International symposium on oilfield chemistry, The Woodlands, Texas, USA, 13–15 April*. https://doi.org/10.2118/173738-MS.)

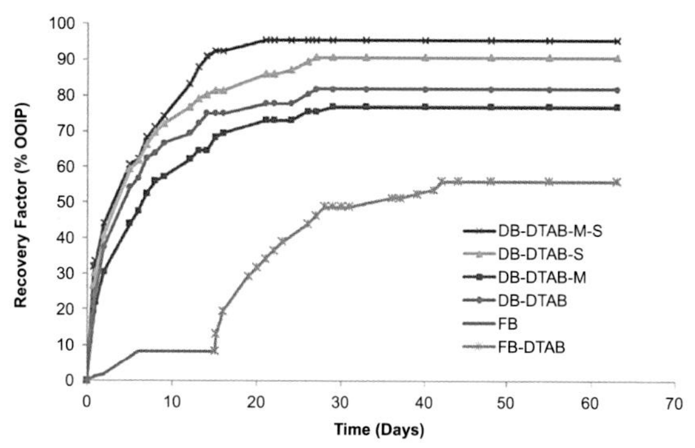

FIG. 4.28 Oil recovery from the spontaneous imbibition test using different imbibing brines in the presence of cationic surfactant. (Credit: From Karimi, M., Al-Maamari, R. S., Ayatollahi, S., & Mehranbod, N. (2016). Wettability alteration and oil recovery by spontaneous imbibition of low salinity brine into carbonates: Impact of Mg^{2+}, SO42− and cationic surfactant. *Journal of Petroleum Science and Engineering, 147*, 560–569. https://doi.org/10.1016/j.petrol.2016.09.015.)

fraction is rarely changed as salinity varies. From the IFT measurements of the microemulsion, the minimum IFT is observed on the order of 2×10^{-3} dyne/cm. The cosolvent addition might introduce some benefits of increasing solubility and viscosity control of microemulsion and modifies the surfactant condition for the specific reservoir. The effects of cosolvent on the phase behavior of anionic and zwitterionic surfactants mixture are investigated. The decrease in the volume of Winsor type III is observed with an increase in cosolvent concentration. In addition, more rapid stabilization of Winsor type III formulation is obtained with an increasing solvent concentration. This study drew some conclusions of application of surfactant mixture. The performance of surfactant mixtures can be effective in either high salinity or low salinity conditions. It addition, the mixing ratio, pH, salinity, and cosolvent determine the performance of surfactant mixture application.

Karimi, Al-Maamari, Ayatollahi, and Mehranbod (2016) investigated the spontaneous imbibition of LSSF using cationic surfactant in carbonates. The study tried to investigate the higher EOR potential of LSSF compared with the conventional surfactant flood. In addition, it attempted to enhance its EOR potential by adjusting the ionic composition of low-salinity surfactant solution. Especially, the roles of Mg^{2+} and $SO_4{}^{2-}$ on the performance of LSSF process are quantified. The crude oil from an Omani carbonate reservoir has TAN of 0.37 mg KOH/g. The synthetic brine of 196,010 ppm TDS is prepared. The low-salinity water is the diluted synthetic brine by a factor of 100. The additional concentration of Mg^{2+} and/or $SO_4{}^{2-}$ is added into the low-salinity brine to prepare the modified version of low-salinity water. The cationic surfactant is used in this study and it has less adsorption to the positively charged carbonate surface. The CMC of the cationic surfactant in deionized water is determined to be 0.4 wt%, and the surfactant concentration of 0.5 wt% is used in the experiments. The results from the IFT measurement, contact angle measurement, and spontaneous imbibition test are analyzed. Although the Indiana limestone core is used for the spontaneous imbibition test, the oil-wet calcite crystal, which is prepared with crude oil aging process, is used for the contact angle measurement. In the contact angle measurement using n-decane, the low-salinity surfactant solution shows the contact angle of 69 degrees. Addition of Mg^{2+} and/or $SO_4{}^{2-}$ results in the more reduction of contact angle. The effect of Mg^{2+} addition is significant compared with that of $SO_4{}^{2-}$. These observations indicate that controlling the ion concentration enhances wettability modification by LSSF. The IFTs between brines and n-decane are measured in the absence and presence of surfactant. It is reported that the addition of Mg^{2+} and/or $SO_4{}^{2-}$ reduces the IFT between low-salinity surfactant solution and n-decane. The spontaneous imbibition test clearly shows the higher oil recovery of LSSF compared with the conventional surfactant flood using the in situ brine (Fig. 4.28). In addition, LSSFs with the addition of Mg^{2+} and $SO_4{}^{2-}$ or only $SO_4{}^{2-}$ enhance the oil production compared with the LSSF (Fig. 4.28). The LSSF with only Mg^{2+} addition produces slightly less oil production compared with the LSSF. This experimental study demonstrated that optimizing the brine composition can enhance the oil production of cationic surfactant flood in carbonates.

Mirchi (2018) published the pore-scale investigation to quantify pore-scale fluid configurations through systematic coreflooding. The systematic coreflooding is developed to be integrated with the micro-CT scanner. The systematic apparatus measures the in situ contact angle and visualizes the pore space and fluid occupancy during the carbonate coreflooding. The experiments use two different brines of high salinity and low salinity. The cationic surfactant of 0.2 wt% is added to the brines, and conventional surfactant solution and low-salinity surfactant solutions are prepared. The IFTs between brines/surfactant solution and crude oil are determined at 500 psi and 80°C. The low-salinity brine shows the lower IFT than the high-salinity brine. When the cationic surfactant is present, the high-salinity brine results in lower IFT than the low-salinity brine. These observations indicate the high salinity condition is close to the optimum salinity of microemulsion system.

The systematic apparatus measures the in situ contact angle during coreflooding and constructs the distribution of in situ contact angle (Fig. 4.29). The distribution indicates the pore-scaled wettability. The distribution of in situ contact angle shows that initial core is estimated to have average in situ contact angle with 140 degrees and wettability is determined to be oil-wet. The injections of high-salinity water and low-salinity water reduce the average in situ contact angle. The LSWF decreases the contact angle more compared with the conventional waterflood injecting high-salinity water. When the surfactant is added in the low-salinity brine, the average in situ contact angle is highly reduced below 100°C. The remaining oil in oil-wet system after waterflood is easily trapped in the corners and crevices. The distributions of the remaining oil are observed before coreflooding and after LSWF or LSSF (Fig. 4.30). The distribution of remaining oil indicates that the LSSF is effective to reduce the trapped oil compared with the LSWF. In addition, the performance of the injection is analyzed according to the pore size. The pore size is categorized with four sections. For the smallest pore size ranging the order of 100–200 μm, the LSSF is still effective to decrease in situ contact angle (Fig. 4.31). The oil extraction from each pore size range is estimated. Major oil production is attributed to the mobilization of oil in largest pore size range. The addition of surfactant enables extraction of more oil from the smallest range of pore size. In line with the results of in situ contact angle measurement, the LSSF recovers oil from the smallest range of pore size section. As a result, the LSSF shows the higher EOR production than LSWF and high-salinity surfactant flood, which is closer to the optimum salinity surfactant. In addition, the oil production can be enhanced when secondary LSWF process is applied ahead of tertiary LSSF. This study visually demonstrated the higher performance of hybrid LSSF displacement by configuring the microscopic fluid distribution and in situ contact angle distribution in oil-wet system. A couple of conclusions are drawn from the study. The LSSF significantly modifies the wettability of oil-wet carbonates and is effective to recover the trapped oil in the corners or crevices and small size of pores. The performance of LSSF can be improved with preflush LSWF.

Teklu et al. (2018) extended the application of LSWF and LSSF to the recovery from liquid-rich Bakken shale reservoir. The spontaneous imbibition test evaluates the shale oil recoveries by LSWF and LSSF processes. The three Bakken shale cores have permeability in the range of 0.001–2.74 md and porosity in the range of 5.29% –7.71%. The Bakken shale crude oil has TAN of 0.09 mg KOH/g and TBN of 1.16 mg KOH/g. The high-salinity water of 240,000 ppm KCl and low-salinity water of 20,000 ppm KCl are synthetically prepared. The anionic surfactant of 1000 ppm is added to the brines. The LSWF shows higher oil recovery by about 14% compared with the high-salinity waterflood. Although the oil recovery during surfactant process is not quantitatively measured, the oil recoveries of the spontaneous imbibition tests are visually investigated. When the low-salinity water is switched to the low-salinity surfactant solution, some oil droplets are expelled from the cores. However, the high-salinity surfactant solution does not show further extrusion of oil droplets from core after high-salinity water. The oil production of LSSF from the spontaneous imbibition test is briefly explained with driving mechanisms including osmosis, capillary pressure, wettability modification, and IFT decrease.

Numerical simulations

Tavassoli, Korrani, Pope, and Sepehrnoori (2016) developed the numerical simulation of LSSF implementing the mechanisms of surfactant flood and LSWF and comprehensive geochemical reactions. The study applied the in-house simulator of UTCHEM-IPhreeqc to simulate the LSWF and LSSF based on the sandstone experimental results of Alagic and Skauge (2010). Because the geochemical reactions are of

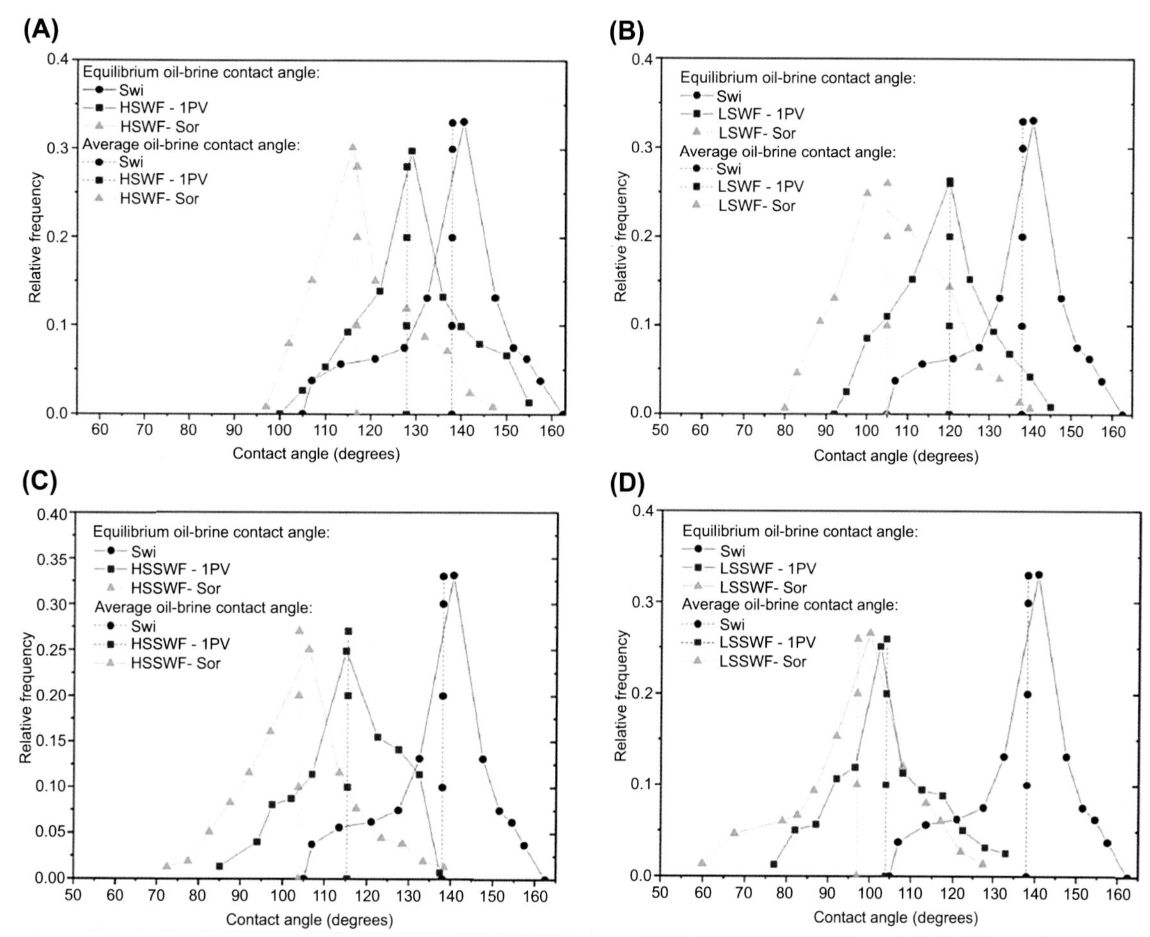

FIG. 4.29 Distribution of in situ contact angle during coreflooding of **(A)** conventional water flood, **(B)** low-salinity waterflood, **(C)** high-salinity surfactant flood, and **(D)** low-salinity surfactant flood. (Credit: From Mirchi, V. (2018). Pore-scale investigation of the effect of surfactant on fluid occupancies during low-salinity waterflooding in carbonates. *Paper presented at the SPE Annual technical Conference and exhibition, Dallas, Texas, USA, 24–26 September*. https://doi.org/10.2118/194045-STU.)

importance in the modeling of LSWF mechanism, the aqueous reactions, mineral reactions of quartz, K-feldspar, and calcite, and partitioning reaction of acidic component of oil between aqueous and oleic phases are incorporated in the modeling. Extensive mechanisms are proposed to explain the experimental observation of LSWF as summarized in Chapter 2. This numerical study adapts the wettability modification as the mechanism of LSWF, which is modeled with the modification of relative permeability and capillary pressure. It is assumed that the wettability modification is attributed to the double layer expansion by the total ionic strength change. Therefore, the total ionic strength is used as the interpolation parameter modifying relative permeability and capillary pressure between high and low salinity threshold conditions. In addition, the IFT reduction and wettability modification in the mechanisms of surfactant EOR also reduce the residual oil saturation and improve the relative permeability. The numerical modeling approach on the mechanisms of surfactant EOR process modifies the residual oil saturation as a function of capillary number and the relative permeability curves as a function the reduced residual oil saturation. Both mechanisms of LSWF and surfactant flood modify the relative permeability curves. The simulation of hybrid LSSF also incorporates the salinity-dependent adsorption of surfactant, microemulsion viscosity, and salinity-dependent IFT reduction using solubilization ratio and Huh's equation (Huh, 1979). The numerical

(A)

(B)

(C)

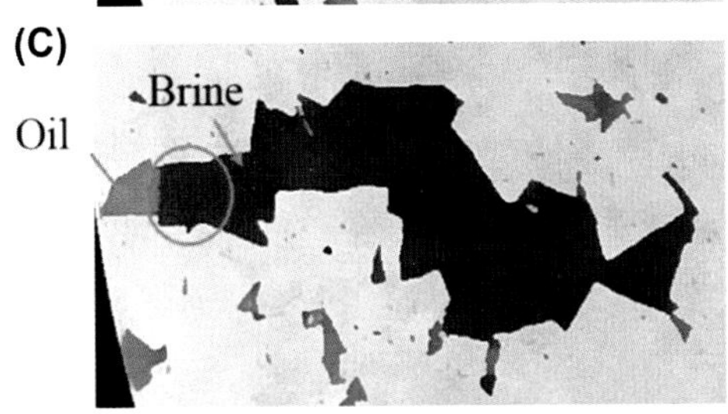

FIG. 4.30 Distribution of remaining oil in oil-wet pore **(A)** before coreflooding and after **(B)** waterflood and **(C)** low-salinity surfactant flood. (Credit: From Mirchi, V. (2018). Pore-scale investigation of the effect of surfactant on fluid occupancies during low-salinity waterflooding in carbonates. *Paper presented at the SPE Annual technical Conference and exhibition, Dallas, Texas, USA, 24–26 September.* https://doi.org/10.2118/194045-STU.)

simulation matches the experimental data of coreflooding including oil recovery, pressure gradient, and effluent ion concentration. Simulations of LSWF and LSSF successfully reproduce the historical data of experiments of LSWF and LSSF.

Dang, Nghiem, Fedutenko, et al. (2018) developed three-dimensional interpolation approach to model the modification of relative permeability considering both mechanisms of LSWF and surfactant flood. The study also incorporated the modeling of comprehensive geochemistry, which is a crucial factor for the LSWF mechanism, and IFT reduction using the solubilization ratio and Huh's equation (Huh, 1979). In comparison with Tavassoli et al. (2016), Dang, Nghiem, Fedutenko, et al. (2018) used the cation exchange, not total ionic strength, on the clay to simulate the wettability modification of LSWF in sandstone reservoirs and IFT reduction to simulate the mechanisms of surfactant. For the simulation of LSWF process, the modification of relative permeability is a function of the cation exchange. For the simulation of hybrid LSSF, another approach of modification of relative permeability is proposed as a function of both cation exchange and IFT reduction (Fig. 4.32). This study successfully validated the numerical models of both LSWF and LSSF using the history-matching process based on the experimental study (Alagic & Skauge, 2010).

ALKALINE FLOOD

Backgrounds of Alkaline Flood

The knowledge of mechanisms and properties of alkaline flood would assist the understanding of low salinity–augmented alkaline flood or low salinity–augmented other chemical EOR. Before the description of low salinity–augmented alkaline flood, the key features of alkaline flood are briefly summarized. Detailed discussion can be found in the references (Lake, 1989; Sheng, 2011). The alkaline flood, i.e., caustic flood, used the in situ surfactant generation and emulsification to enhance oil recovery and also supports the other chemical EOR processes such as polymer flood and/or surfactant flood. The chemicals such as sodium carbonate (Na_2CO_3) and sodium hydroxide (NaOH) are the common alkali agents. Other types of alkali chemicals and organic alkali are also used to avoid precipitation problem. In alkali flood, the chemical reaction between alkali chemicals and organic acids of crude oil generates in situ surfactants, i.e., soap, reducing IFT. The common alkali agents of NaOH and Na_2CO_3 dissociate, respectively. Although both alkali agents yield OH^- controlling pH, Na_2CO_3 requires the hydrolysis reactions Eqs. (4.19–4.21).

$$NaOH \rightleftarrows Na^+ + OH^- \tag{4.19}$$

$$Na_2CO_3 \rightleftarrows 2Na^+ + CO_3^{2-} \tag{4.20}$$

$$CO_3^{2-} + H_2O \rightleftarrows HCO_3^- + OH^- \tag{4.21}$$

The alkaline reacts with the crude oil formulating the in situ soap and emulsification. In addition, the ionic strength and pH of solution influence the reactions to form the in situ soap and emulsification.

Firstly, the in situ soap generation is described. When the injected alkali chemicals contact crude oil, saponifiable components, i.e., naphthenic acids, of crude oil react with the alkali component. The majority of naphthenic acids represent the mixture of cyclopentyl and cyclohexyl carboxylic acids, which approximately have molecular weight of 120 to well over 700. The simplified form of alkali-crude oil chemistry is

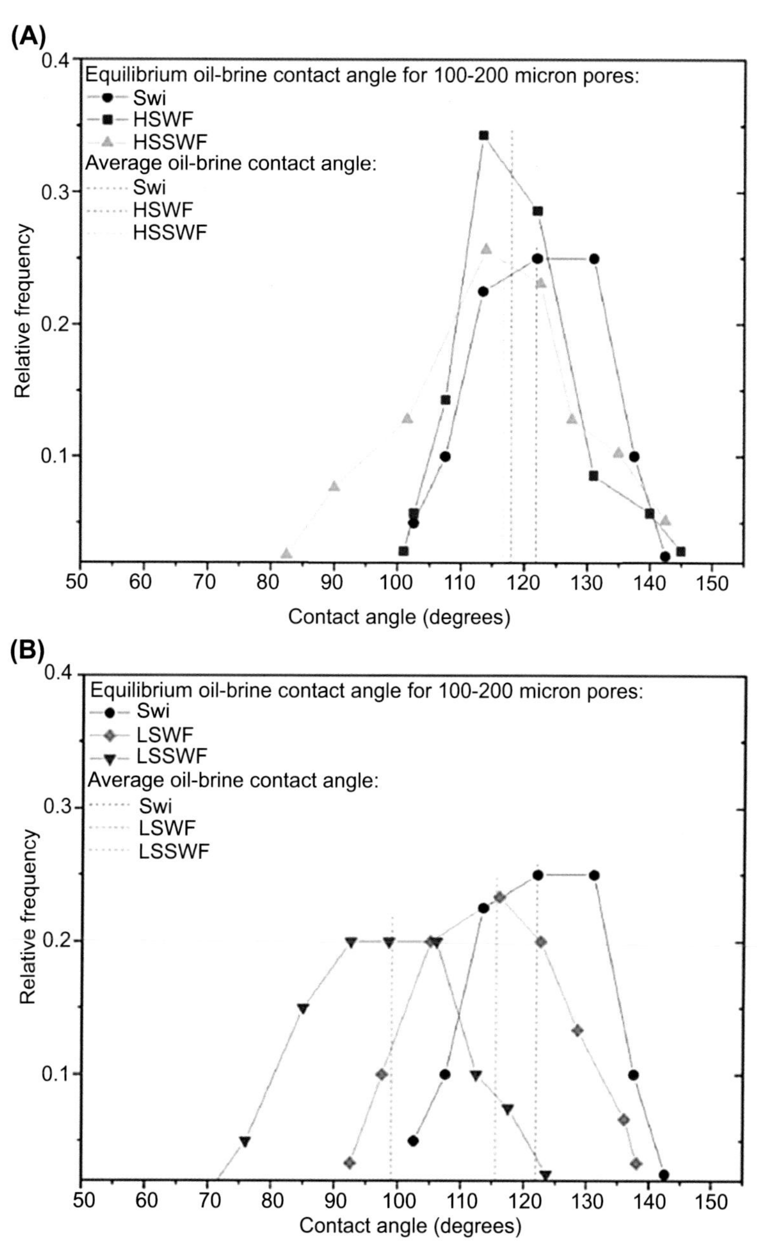

FIG. 4.31 Distribution of in situ contact angle during coreflooding of **(A)** conventional waterflood and high-salinity surfactant flood, and **(B)** low-salinity waterflood and low-salinity surfactant flood in the pore size on the order of 100–200 μm. (Credit: From Mirchi, V. (2018). Pore-scale investigation of the effect of surfactant on fluid occupancies during low-salinity waterflooding in carbonates. *Paper presented at the SPE Annual technical Conference and exhibition, Dallas, Texas, USA, 24–26 September*. https://doi.org/10.2118/194045-STU.)

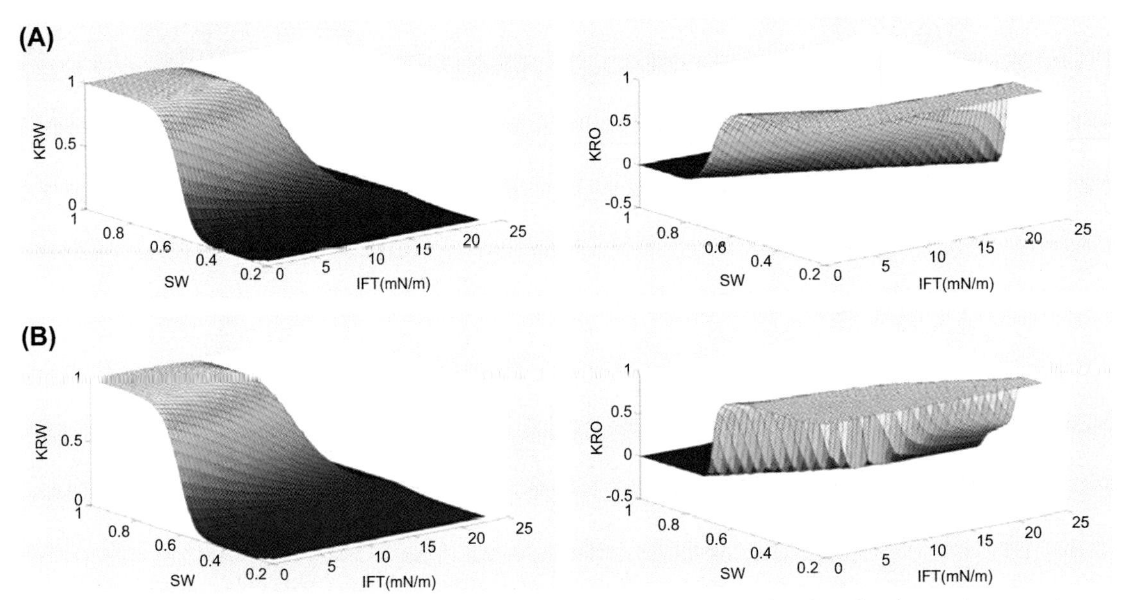

FIG. 4.32 The three-dimensional modification of relative permeability as a function of cation exchange and IFT reduction in **(A)** low-salinity water condition and **(B)** high-salinity water condition. (Credit: From Dang, C., Nghiem, L., Fedutenko, E., Gorucu, E., Yang, C., & Mirzabozorg, A. (2018a). Application of artificial intelligence for mechanistic modeling and probabilistic forecasting of hybrid low salinity chemical flooding. *Paper presented at the SPE Annual technical conference and exhibition, Dallas, Texas, USA, 24–26 September.* https://doi.org/10.2118/191474-MS.)

described by a couple of reactions. The partitioning reaction distributes the organic acids between aqueous and oleic phases. The subsequent hydrolysis reaction produces a soluble anionic surfactant, which is conventionally described as carboxylate, $RCOO^-$, in the presence of alkali additive. deZabala, Vislocky, Rubin, and Radke (1982) schematically depict the reactions (Fig. 4.33). The overall reaction of sodium hydroxide alkali is described in Eq. (4.22).

$$HA_o + NaOH \rightleftarrows NaA + H_2O \tag{4.22}$$

where HA_o indicates the organic acid component of crude oil and A indicates the organic anionic surfactant.

The reaction occurs in the interface between aqueous and oleic phases. Recalling the partitioning coefficient of Eqs. (3.69) and (3.70), the organic acid components are distributed between aqueous and oleic phases. The organic acid components in aqueous phase undergo the hydrolysis reaction following Eqs. (3.71) and (3.72). With the dissociation of water of Eq. (4.23) and Eqs. (3.69) and (3.72), the concentration of anionic surfactant is calculated with Eq. (4.24). It is noticeable that the concentration of anionic surfactant is controlled by pH. The surface-active agent, anionic surfactant component, is accumulated at the interface of oil and water, and then it reduces the IFT. The IFT reduction introduces the wettability modification as well as emulsification.

$$K_w = [H^+][OH^-] \tag{4.23}$$

$$[A^-] = \frac{K_D K_a [HA_o][OH^-]}{K_w} \tag{4.24}$$

where K_w is a dissociation constant of water.

Low IFT by the in situ soap generation leads to formulating the emulsification. Because the stability of emulsion depends on the properties of oil and water interface, the IFT reduction by the acidic component of crude oil easily makes the emulsion. The crude oil with negligible organic acidic component is hardly emulsified with alkali agents because of high IFT. The addition of surfactant reduces the IFT of the interface, and then emulsification occurs. The emulsification can improve the sweep efficiency by blocking the smaller pore throat or contribute to the continuous oil bank in porous media.

The alkaline injection has potentials including precipitation, adherence to rock surface, and reactions with mineral and water. Some alkalis such as NaOH

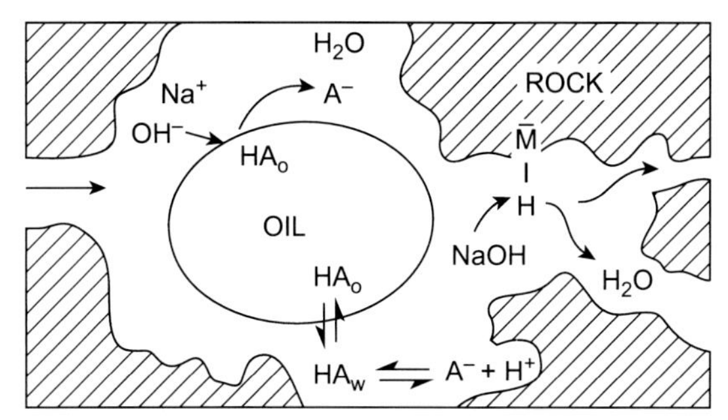

FIG. 4.33 The schematic description of reactions of alkaline recovery process. (Credit: From deZabala, E. F., Vislocky, J. M., Rubin, E., & Radke, C. J. (1982). A chemical theory for linear alkaline flooding. *SPE Journal, 22*(02), 245–258. https://doi.org/10.2118/8997-PA.)

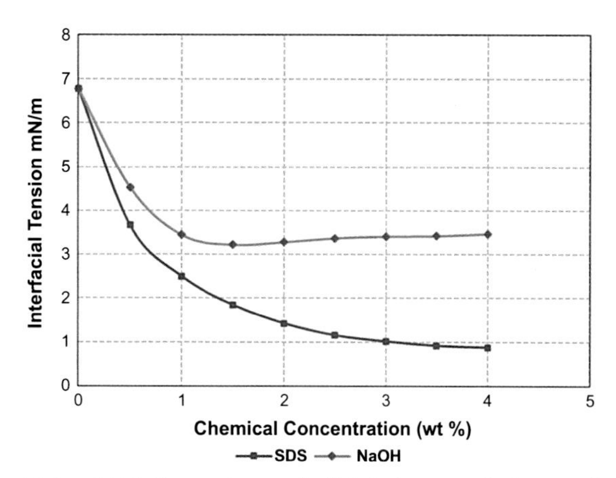

FIG. 4.34 Comparison of IFT between alkaline and surfactant solution in low salinity condition. (Credit: From Shaddel, S., & Tabatabae-Nejad, S. A. (2015). Alkali/surfactant improved low-salinity waterflooding. *Transport in Porous Media, 106*(3), 621–642. https://doi.org/10.1007/s11242-014-0417-1.)

and Na_2CO_3 have a risk to precipitate with divalent cations. In carbonate reservoirs, the presence of anhydrite, $CaSO_4$, or gypsum, $CaSO_4 \cdot 2H_2O$, can make precipitation when NaOH or Na_2CO_3 is added. Higher content of divalent cations in formation water also leads to the precipitation during the alkaline injection. To avoid the precipitation problem, the agents of sodium bicarbonate ($NaHCO_3$) and sodium sulfate (Na_2SO_4) are added with alkali agents. In addition, alkali has attraction to the reservoir rock. Similar to the description of cation exchange in previous chapter, the alkaline undergoes ion-exchange reactions. Another significant reaction is the direct reaction between specific mineral and alkali agent. The alkali agent has reactivity with a number of minerals and results in the dissolution or precipitation.

Low Salinity–augmented Alkaline Flood

Because higher pH increases the ionic strength and the effects of ionic strength or salinity on the performance of alkaline flood are relatively less predominant, a few studies have investigated the hybrid process of low salinity–augmented alkaline flood (LSAF). Shaddel and Tabatabae-Nejad (2015) designed the hybrid process of LSSF and LSAF and evaluated the synergetic effects of LSAF compared with that of LSSF. The alkalis of NaOH and anionic surfactant are prepared. The evaluations ignore the use of cosurfactant or cosolvent. The formation brine has 209,853 ppm TDS, and the dilution of the formation brine by factors of 10 and 100 makes the low-salinity brines. The two different crude oils are used for the experiments. IFT measurement and coreflooding illustrate the performance of LSSF and LSAF. The slight reduction in IFT is observed with a decrease in salinity. Additions of surfactant and alkali chemicals significantly reduce the IFT, and surfactant addition shows slightly higher reduction compared with the alkali addition (Fig. 4.34). Although the surfactant has higher potential to reduce IFT than alkali, it is clearly observed that in situ surfactant generation from the interaction between alkali and crude oil also decreases IFT and potentially recovers the trapped oil. The experiments of coreflooding compare the tertiary recovery of LSSF and LSAF after secondary conventional waterflood. The LSSF and LSAF enhance the oil recovery by 22% and 12.3%, respectively. In addition, the overall pressure of injection is higher in LSSF and LSAF compared with the LSWF. Because the preliminary study estimates less swelling of clay in the low-salinity water condition, it is explained that higher pressure is attributed to the release of extra oil and slightly higher viscosity of chemical solutions during LSSF and LSAF. The study drew a couple of conclusions regarding the LSAF. Although the LSAF shows less EOR potential than LSSF, the lower cost of alkali agent has benefits on the EOR implementation. In addition to the reducing chemical cost, the alkali agent injection could prevent the adsorption of in situ generated petroleum surfactant.

The studies (Suleimanov, Latifov, Veliyev, & Frampton, 2017, 2018) have investigated the low-salinity and low-hardness alkali as the EOR process in terms of IFT, contact angle, emulsion stability, adsorption, clay swelling using bentonite, and secondary and tertiary recoveries of coreflooding. The crude oil has TAN of 0.15 mg KOH/g and TBN of 1.29 mg KOH/g. The formation water of 231,194.9 ppm TDS and synthetic water of 14,063.3 ppm TDS are prepared. The synthetic water is modified to make the low-salinity and low-hardness alkaline solution by extracting the divalent cations and adding the chemical additive. Two samples of low-salinity and low-hardness alkali solutions with different concentrations of chemical additive are prepared. The diluted synthetic water of 3587.45 ppm TDS is also investigated. As the salinity decreases and pH increases, both IFT and contact angle decrease. The results of IFT and contact angle measurements indicate that the acidic component of crude oil reduces the IFT at high pH condition and modifies the wettability. In the adsorption test, the low-salinity and low-hardness alkali solution shows slightly higher adsorption than the synthetic seawater and low-salinity water, but the discrepancy is acceptable. The stability test of emulsion measures the separation efficiency of water from emulsion system for the low-salinity water and low-salinity and low-hardness solutions. Lower separation efficiency is observed for the alkali solutions compared with the low-salinity water. It implies that the low-salinity and low-hardness alkali solution generates more stable water-in-oil emulsion. In addition, it results in less swelling of bentonite. With the favorable aspects of low-salinity and low-hardness alkali solution, higher oil recovery is obtained in displacement experiments compared with the waterflood and LSWF.

ALKALINE-SURFACTANT-POLYMER FLOOD

The chemical EOR has a huge potential to employ the synergetic effects by incorporating the low-salinity water or smart water as the makeup brine. Although the mechanism and effect of LSWF are not fully clarified, the LSWF obviously plays positive roles on the stability of chemical, adsorption of chemical, formation damage, optimum condition for the chemical EOR, etc. With the clear evidence of the synergy between LSWF and single chemical EOR process, the coapplications of alkaline, surfactant, and polymer flood to be combined with the LSWF are further investigated to enhance the synergy.

Previously, Johannessen and Spildo (2014) reported an enhanced potential to reduce IFT by the LSSF and the higher oil recovery from sandstone coreflooding with the comparable capillary number improvement. In the coreflooding experiment, polymer slug injection is applied to support the LSSF. The injection of Flopaam 3630s polymer slug follows the surfactant slug injection. The coreflooding experiments are designed with secondary waterflood, and tertiary surfactant flood followed by polymer flood. The coreflooding experiments use high- and low-salinity makeup brines. As observed in the previous studies of LSPF, the polymer has less degradation in low salinity condition to reach the target viscosity. Less polymer concentration by a factor of 2 is required for the low-salinity polymeric solution compared with the high-salinity polymeric solution. The polymer slug injection increases the differential pressure regardless of salinity condition. Despite the less polymer concentration, significantly higher increment of differential pressure is observed in LSPF compared with the polymer flood using high salinity. It implies the higher injectivity to be secured for LSPF. Before the polymer slug injection, the surfactant slug size of 0.5 PV is designed to be injected. During the surfactant injection with small slug size, oil recovery negligibly or hardly increases. However, the succeeding polymer flood produces immobile oil because of the preinjected surfactant additive (Fig. 4.35) and reduces 717% of residual oil saturation. It is obviously concluded that low salinity—augmented surfactant/polymer flood has enough potential to produce the trapped oil by improving both displacement and sweep efficiencies. Synergetic effects between LSPF and LSSF can be secured through low salinity—augmented surfactant/polymer flood (LSSP) as well as low chemical injection.

Wang, Ayirala, AlSofi, Al-Yousef, and Aramco (2018) published the experimental attempts of LSSF for the carbonate oil recovery. The connate water has salinity of 213.723 mg/L TDS, and the high-salinity water has salinity of 57,610 mg/L TDS. The low-salinity water is prepared by diluting the high-salinity water by a factor of 10. A sulfonated polyacrylamide and betaine-type amphoteric surfactant are used. The contact angle and ζ-potential are measured to confirm the potential of wettability modification in low-salinity water condition for the specific carbonate rock. The reduction in contact angle and decrease toward more negative ζ-potential are observed in low-salinity water condition compared with the high-salinity water condition. These observations consistently agree with the previous experimental results of LSWF. It is evaluated that the LSWF modifies the wettability of carbonate rock from oil-wet to weakly oil-wet or intermediate wet. With the evidence of LSWF effectiveness, the tertiary LSSP application is evaluated and compared with the conventional surfactant-polymer flood using high-salinity water as makeup

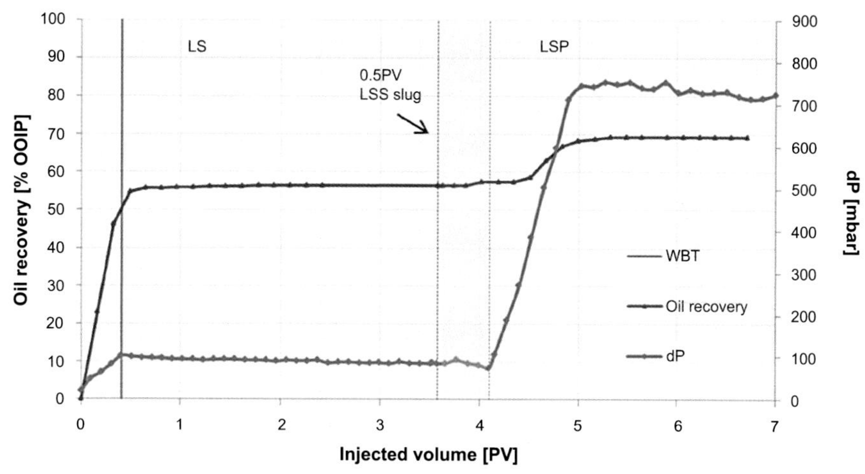

FIG. 4.35 The oil recovery and differential pressure of coreflooding of low salinity–augmented surfactant/polymer floods. (Credit: From Johannessen, A. M., & Spildo, K. (2014). Can lowering the injection brine salinity further increase oil recovery by surfactant injection under otherwise similar conditions? *Energy and Fuels*, *28*(11), 6723–6734. https://doi.org/10.1021/ef500995y.)

brine. The LSSP is designed with the coinjection of polymer and surfactant additives. The polymer concentration to be injected in LSSP halves but has the 70% of target viscosity compared with the chemical solution of conventional surfactant-polymer flood. In terms of surfactant, the equal concentration of surfactant is used for both low- and high-salinity water conditions, and comparable IFT reduction is observed. The hybrid LSSP shows the higher oil cut and differential pressure despite less polymer injection. This study also demonstrated that the hybrid LSSP improves both sweep and displacement efficiencies and secures the practicality and economics of chemical EOR process.

The hybrid alkaline/surfactant/polymer flood (ASP) has been widely investigated because of a number of reasons. This hybrid process improves both sweep and displacement efficiencies and highly enhances the oil production. The alkali addition produces the in situ surfactant generation, which means the less injection of expensive surfactant. It also reduces the adsorption of surfactant on the rock surface at high pH condition and secures the stability of surfactant by reacting with the divalent cations. The application of low-salinity water as makeup brine into the hybrid ASP process is expected to employ the additional synergy of low-salinity water and chemical additives, as discussed in previous sections, and enhance wettability modification into ASP process.

Because the mechanisms of ASP processes are sensitive to the salinity, pH, etc., a number of studies have a primary object to quantify the effects of geochemistry on the synergetic effects and mechanisms of the hybrid ASP process. The studies (Dang, Nghiem, Nguyen, et al. 2018; Farajzadeh, Matsuura, Batenburg, & Harm, 2012; Ghadami, Deo Tewari, Zarubinska, & Motaei, 2018; Hosseini-Nasab, Padalkar, Battistutta, & Zitha, 2016; Nghiem, Skoreyko, Gorucu, Dang, & Shrivastava, 2017) have developed the mechanistic simulation of ASP process to be coupled with fully geochemical reactions. Based on the comprehensive studies of ASP process, the application of low-salinity water/smart water as the makeup brine is implemented into the hybrid ASP process. Reducing the salinity, ionic strength might be optimistic for polymer flood but might not be for surfactant and alkaline floods owing to optimum salinity condition. Reduction in hardness could bring the overall positive benefits to chemical EOR. Therefore, many studies have investigated hybrid ASP process optimizing the salinity, hardness, ionic composition, etc.

Battistutta, van Kuijk, Groen, and Zitha (2015) conducted the comprehensive experiments of ASP process. The study estimated the optimal salinity to make the Winsor type III of microemulsion as the 1.5 wt% NaCl. At the optimal salinity, the ultralow IFT is obtained from the IFT measurement and phase behavior test. A variety of ASP formulations are designed. The makeup brines of the ASP formulations have the salinities of 0.5, 1, 1.5, and 2.0 wt% NaCl. Because the optimal salinity is determined as 2.0 wt%, the ASP formulations having the salinity less than 2.0 wt% are at underoptimum salinity condition representing Winsor type I of microemulsion system. In the coreflooding using sandstone core, the performance of the various ASP formulations is investigated in terms of aqueous solubility of microemulsion, oil recovery, and retention of surfactant and polymer. The experiments observed that the underoptimum salinity condition brings advantages to the ASP process. The aqueous solubility of microemulsion is secured without the cosurfactant addition. Higher solubility produces the clean oil bank recovery as well as reducing cost of oil treatment at production facilities. In addition, the underoptimum salinity results in less retention of chemical additives and can avoid the risk to undergo the overoptimum salinity condition. Hosseini-Nasab et al. (2016) developed the mechanistic model of ASP process under suboptimum salinity conditions based on the experimental results of Battistutta et al. (2015). The mechanistic simulation of ASP process numerically models the polymer rheology, phase behavior of microemulsion, adsorption of chemicals, wettability modification, and reduction of residual oil saturation considering the mechanisms of ASP process, and mechanical and chemical interactions in crude oil/brine/rock systems. The historical results of oil production, differential pressure, pH, and effluent concentrations are accurately reproduced through simulations of underoptimum salinity–augmented ASP process.

Dang, Nghiem, Nguyen, et al. (2018) also simulated the ASP process and evaluated the hypothetical application of low salinity–augmented ASP process. The main difference between this study and work of Hosseini-Nasab et al. (2016) is the modeling of LSWF mechanism. The previous study of Hosseini-Nasab et al. (2016) ignored the mechanism of LSWF because the suboptimum condition of 1.5 wt% is much higher than the optimal condition, which is known as less than 5000 ppm for sandstone, to introduce the mechanism of LSWF. Instead, the study mainly focused on the wettability modification following the mechanisms of alkaline/surfactant flood. Dang, Nghiem, Nguyen, et al. (2018) developed the mechanistic model of ASP process and applied the mechanism of LSWF, wettability modification, to simulate the low salinity–augmented ASP process. The mechanistic model of ASP process covers the polymer rheology, phase behavior of microemulsion, adsorption of chemicals, relative permeability modification by the wettability modification, and reduction of residual oil saturation. For the modeling of

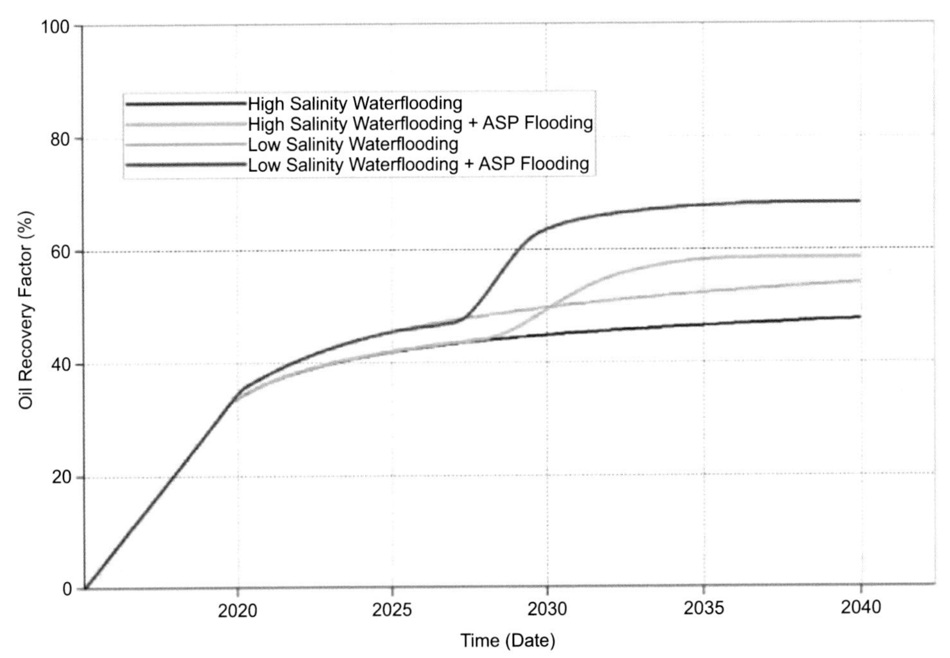

FIG. 4.36 The oil recovery of low salinity–augmented ASP flood compared with the low salinity waterflood and conventional ASP flood. (Credit: From Dang, C., Nghiem, L., Nguyen, N., Yang, C., Chen, Z., & Bae, W. (2018b). Modeling and optimization of alkaline-surfactant-polymer flooding and hybrid enhanced oil recovery processes. *Journal of Petroleum Science and Engineering, 169*, 578–601. https://doi.org/10.1016/j.petrol.2018.06.017.)

wettability modification, relative permeability is modified by the ion-exchange reactions following one of the suggested mechanisms of LSWF. The simulations compare the performance of low salinity–augmented ASP process to that of conventional ASP process using high-salinity water (Fig. 4.36) and optimize the injection designs of low salinity–augmented ASP processes to maximize net present value (NPV) and oil production. Other studies (Farajzadeh et al. 2012; Ghadami et al. 2018; Nghiem et al. 2017) have simulated the ASP process following the similar mechanistic approach of numerical modeling.

Because the chemical EOR process requires the expensive chemical injection costs, optimum condition is of importance to the successful EOR application. The optimum condition varies according to the designs of chemical EOR process. The reducing salinity and/or hardness is favorable to the chemical process injecting polymer. Only reducing hardness can be the optimistic condition for process with surfactant and/or alkali injections. In spite of the same chemical EOR process, the optimistic condition can vary with mineralogy, type of chemical additive, formation brine composition, etc. The application of low-salinity water/smart water does not guarantee the optimistic condition of chemical EOR formulation. It is necessary to investigate the detailed and comprehensive investigations of chemical EOR using low-salinity water flood/smart waterflood to optimize the potential of hybrid EOR. The referred studies have demonstrated the high oil recovery and synergy of low salinity–augmented chemical EOR. It is obvious that the usage of low-salinity water and smart water extensively introduces the technical and economic benefits to the extensive applications of chemical EOR process.

REFERENCES

Alagic, E., & Skauge, A. (2010). Combined low salinity brine injection and surfactant flooding in mixed–wet sandstone cores. *Energy and Fuels, 24*(6), 3551–3559. https://doi.org/10.1021/ef1000908.

Alagic, E., Spildo, K., Arne, S., & Jonas, S. (2011). Effect of crude oil ageing on low salinity and low salinity surfactant flooding. *Journal of Petroleum Science and Engineering, 78*(2), 220–227. https://doi.org/10.1016/j.petrol.2011.06.021.

Alhuraishawy, A. K., Abdulmohsin Imqam, Wei, M., & Bai, B. (2016). Coupling low salinity water flooding and preformed particle gel to enhance oil recovery for fractured carbonate reservoirs. In *Paper presented at the SPE western regional meeting, Anchorage, Alaska, USA, 23–26 May*. https://doi.org/10.2118/180386-MS.

Alhuraishawy, A. K., & Bai, B. (2017). Evaluation of combined low-salinity water and microgel treatments to improve oil recovery using partial fractured carbonate models. *Journal of Petroleum Science and Engineering, 158*, 80–91. https://doi.org/10.1016/j.petrol.2017.07.016.

Alhuraishawy, A. K., Bai, B., & Mingzhen, W. (2018). Combined ionically modified seawater and microgels to improve oil recovery in fractured carbonate reservoirs. *Journal of Petroleum Science and Engineering, 162*, 434–445. https://doi.org/10.1016/j.petrol.2017.12.052.

Almansour, A. O., AlQuraishi, A. A., AlHussinan, S. N., & AlYami, H. Q. (2017). Efficiency of enhanced oil recovery using polymer-augmented low salinity flooding. *Journal of Petroleum Exploration and Production Technology, 7*(4), 1149–1158. https://doi.org/10.1007/s13202-017-0331-5.

AlSofi, A. M., Wang, J., AlBoqmi, A. M., AlOtaibi, M. B., Ayirala, S. C., & AlYousef, A. A. (2016). SmartWater synergy with chemical EOR for a slightly viscous Arabian heavy reservoir. In *Paper presented at the SPE heavy oil conference and exhibition, Kuwait City, Kuwait, 6–8 December*. https://doi.org/10.2118/184163-MS.

AlSofi, A. M., Wang, J., AlBoqmi, A. M., AlOtaibi, M. B., Ayirala, S. C., & AlYousef, A. A. (2018a). SmartWater synergy with chemical enhanced oil recovery: Polymer effects on smartwater. *SPE Reservoir Evaluation and Engineering Preprint (Preprint), 17*. https://doi.org/10.2118/184163-PA.

AlSofi, A. M., Wang, J., & Kaidar, Z. F. (2018b). SmartWater synergy with chemical EOR: Effects on polymer injectivity, retention and acceleration. *Journal of Petroleum Science and Engineering, 166*, 274–282. https://doi.org/10.1016/j.petrol.2018.02.036.

Ashraf, A., Hadia, N., Torsaeter, O., & Medad Twimukye Tweheyo. (2010). Laboratory investigation of low salinity waterflooding as secondary recovery process: Effect of wettability. In *Paper presented at the SPE oil and gas India conference and exhibition, Mumbai, India, 20–22 January*. https://doi.org/10.2118/129012-MS.

Ayirala, S. C., Ernesto, U.-N., Matzakos, A. N., Chin, R. W., Doe, P. H., & van den Hoek, P. J. (2010). A designer water process for offshore low salinity and polymer flooding applications. In *Paper presented at the SPE improved oil recovery symposium, Tulsa, Oklahoma, USA, 24–28 April*. https://doi.org/10.2118/129926-MS.

Battistutta, E., van Kuijk, S. R., Groen, K. V., & Zitha, P. L. J. (2015). Alkaline-surfactant-polymer (ASP) flooding of crude oil at under-optimum salinity conditions. In *Paper presented at the SPE Asia pacific enhanced oil recovery conference, Kuala Lumpur, Malaysia, 11–13 August*. https://doi.org/10.2118/174666-MS.

Bird, R. B. (1960). *Transport phenomena*. New York: Wiley.

Bird, R. B., Armstrong, R. C., & Hassage, O. (1987). *Dynamics of polymeric liquids, volume 1: Fluid mechanics* (2nd ed.). Wiley.

Bourrel, M., & Schechter, R. S. (1988). *Microemulsions and related systems: Formulation, solvency, and physical properties, surfactant science series*. New York: M. Dekker.

Brattekås, B., Graue, A., & Seright, R. (2016). Low-salinity chase waterfloods improve performance of Cr(III)-Acetate hydrolyzed polyacrylamide gel in fractured cores. *SPE Reservoir Evaluation and Engineering, 19*(02), 331–339. https://doi.org/10.2118/173749-PA.

Brattekås, B., & Seright, R. S. (2018). Implications for improved polymer gel conformance control during low-salinity chase-floods in fractured carbonates. *Journal of Petroleum Science and Engineering, 163*, 661–670. https://doi.org/10.1016/j.petrol.2017.10.033.

Carreau, P. J. (1972). Rheological equations from molecular network theories. *Transactions of the Society of Rheology, 16*(1), 99–127. https://doi.org/10.1122/1.549276.

Chauveteau, G. (1982). Rodlike polymer solution flow through fine pores: Influence of pore size on rheological behavior. *Journal of Rheology, 26*(2), 111–142. https://doi.org/10.1122/1.549660.

Clarke, A., Howe, A. M., Mitchell, J., Staniland, J., & Hawkes, L. A. (2016). How viscoelastic-polymer flooding enhances displacement efficiency. *SPE Journal, 21*(03), 675–687. https://doi.org/10.2118/174654-PA.

Dang, C., Nghiem, L., Fedutenko, E., Gorucu, E., Yang, C., & Mirzabozorg, A. (2018a). Application of artificial intelligence for mechanistic modeling and probabilistic forecasting of hybrid low salinity chemical flooding. In *Paper presented at the SPE annual technical conference and exhibition, Dallas, Texas, USA, 24–26 September*. https://doi.org/10.2118/191474-MS.

Dang, C., Nghiem, L., Nguyen, N., Yang, C., Chen, Z., & Bae, W. (2018b). Modeling and optimization of alkaline-surfactant-polymer flooding and hybrid enhanced oil recovery processes. *Journal of Petroleum Science and Engineering, 169*, 578–601. https://doi.org/10.1016/j.petrol.2018.06.017.

Davies, J. T. (1957). A quantitative kinetic theory of emulsion type I. Physical chemistry of the emulsifying. In *International congress of surface acitivity*. London: Butterworths.

Delshad, M., Kim, D. H., Magbagbeola, O. A., Huh, C., Pope, G. A., & Tarahhom, F. (2008). Mechanistic interpretation and utilization of viscoelastic behavior of polymer solutions for improved polymer-flood efficiency. In *Paper presented at the SPE symposium on improved oil recovery, Tulsa, Oklahoma, USA, 20–23 April*. https://doi.org/10.2118/113620-MS.

deZabala, E. F., Vislocky, J. M., Rubin, E., & Radke, C. J. (1982). A chemical theory for linear alkaline flooding. *SPE Journal, 22*(02), 245–258. https://doi.org/10.2118/8997-PA.

Farajzadeh, R., Matsuura, T., Batenburg, D.van, & Harm, D. (2012). Detailed modeling of the alkali/surfactant/polymer (ASP) process by coupling a multipurpose reservoir simulator to the chemistry package PHREEQC. *SPE Reservoir Evaluation and Engineering, 15*(04), 423–435. https://doi.org/10.2118/143671-PA.

Ghadami, N., Deo Tewari, R., Zarubinska, M., & Motaei, E. (2018). The impact of ion exchange and surfactant partitioning on ASP modeling, a Brown offshore field. In *Paper presented at the offshore technology conference Asia, Kuala Lumpur, Malaysia, 20–23 March*. https://doi.org/10.4043/28211-MS.

Graiffin, W. C. (1954). Calculation of HLB values of non-ionic surfactants. *Journal of Cosmetic Science, 5*(4), 249–256.

Griffin, W. C. (1949). Classification of surface-active agents by "HLB". *Journal of Cosmetic Science, 1*(5), 311–326.

Gupta, S. P., & Greenkorn, R. A. (1974). Determination of dispersion and nonlinear adsorption parameters for flow in porous media. *Water Resources Research, 10*(4), 839–846. https://doi.org/10.1029/WR010i004p00839.

Hand, D. B. (1929). Dineric distribution. *The Journal of Physical Chemistry, 34*(9), 1961–2000. https://doi.org/10.1021/j150315a009.

Healy, R. N., Reed, R. L., & Stenmark, D. C. (1976). Multiphase microemulsion systems. *SPE Journal, 16*(03), 147–160. https://doi.org/10.2118/5565-PA.

Hosseini-Nasab, S. M., Padalkar, C., Battistutta, E., & Zitha, P. L. J. (2016). Mechanistic modeling of the alkaline/surfactant/polymer flooding process under sub-optimum salinity conditions for enhanced oil recovery. *Industrial and Engineering Chemistry Research, 55*(24), 6875–6888. https://doi.org/10.1021/acs.iecr.6b01094.

Hosseinzade Khanamiri, H., Baltzersen Enge, I., Nourani, M., Stensen, J.Å., Torsæter, O., & Hadia, N. (2016). EOR by low salinity water and surfactant at low concentration: Impact of injection and in situ brine composition. *Energy and Fuels, 30*(4), 2705–2713. https://doi.org/10.1021/acs.energyfuels.5b02899.

Hosseinzade Khanamiri, H., Nourani, M., Tichelkamp, T., Stensen, J.Å., Øye, G., & Torsæter, O. (2016). Low-salinity-surfactant enhanced oil recovery (EOR) with a new surfactant blend: Effect of calcium cations. *Energy and Fuels, 30*(2), 984–991. https://doi.org/10.1021/acs.energyfuels.5b02848.

Huggins, M. L. (1942). The viscosity of dilute solutions of long-chain molecules. IV. Dependence on concentration. *Journal of the American Chemical Society, 64*(11), 2716–2718. https://doi.org/10.1021/ja01263a056.

Huh, C. (1979). Interfacial tensions and solubilizing ability of a microemulsion phase that coexists with oil and brine. *Journal of Colloid and Interface Science, 71*(2), 408–426. https://doi.org/10.1016/0021-9797(79)90249-2.

Johannessen, A. M., & Spildo, K. (2013). Enhanced oil recovery (EOR) by combining surfactant with low salinity injection. *Energy and Fuels, 27*(10), 5738–5749. https://doi.org/10.1021/ef400596b.

Johannessen, A. M., & Spildo, K. (2014). Can lowering the injection brine salinity further increase oil recovery by surfactant injection under otherwise similar conditions? *Energy and Fuels, 28*(11), 6723–6734. https://doi.org/10.1021/ef500995y.

Karimi, M., Al-Maamari, R. S., Ayatollahi, S., & Mehranbod, N. (2016). Wettability alteration and oil recovery by spontaneous imbibition of low salinity brine into carbonates: Impact of Mg^{2+}, SO_4^{2-} and cationic surfactant. *Journal of Petroleum Science and Engineering, 147*, 560–569. https://doi.org/10.1016/j.petrol.2016.09.015.

Khanamiri, H. H., Torsæter, O., & Stensen, J.Å. (2015). Experimental study of low salinity and optimal salinity surfactant injection. In *Paper presented at the EUROPEC 2015, Madrid, Spain, 1–4 June*. https://doi.org/10.2118/174367-MS.

Khorsandi, S., Qiao, C., & Johns, R. T. (2017). Displacement efficiency for low-salinity polymer flooding including wettability alteration. *SPE Journal, 22*(02), 417–430. https://doi.org/10.2118/179695-PA.

Kim, D. H., Lee, S., Ahn, C. H., Huh, C., & Pope, G. A. (2010). Development of a viscoelastic property database for EOR polymers. In *Paper presented at the SPE improved oil recovery symposium, Tulsa, Oklahoma, USA, 24–28 April*. https://doi.org/10.2118/129971-MS.

Kraemer, E. O. (1938). Molecular weights of celluloses and cellulose derivates. *Industrial & Engineering Chemistry, 30*(10), 1200–1203. https://doi.org/10.1021/ie50346a023.

Lake, L. W. (1989). *Enhanced oil recovery*. Englewood Cliffs, N.J: Prentice Hall.

Lee, S., Kim, D. H., Huh, C., & Pope, G. A. (2009). Development of a comprehensive rheological property database for EOR polymers. In *Paper presented at the SPE annual technical conference and exhibition, New Orleans, Louisiana, 4–7 October*. https://doi.org/10.2118/124798-MS.

Mirchi, V. (2018). Pore-scale investigation of the effect of surfactant on fluid occupancies during low-salinity waterflooding in carbonates. In *Paper presented at the SPE annual technical conference and exhibition, Dallas, Texas, USA, 24–26 September*. https://doi.org/10.2118/194045-STU.

Mohammadi, H., & Jerauld, G. (2012). Mechanistic modeling of the benefit of combining polymer with low salinity water for enhanced oil recovery. In *Paper presented at the SPE improved oil recovery symposium, Tulsa, Oklahoma, USA, 14–18 April*. https://doi.org/10.2118/153161-MS.

Nelson, R. C. (1982). The salinity-requirement diagram – a useful tool in chemical flooding research and development. *SPE Journal, 22*(02), 259–270. https://doi.org/10.2118/8824-PA.

Nghiem, L., Skoreyko, F., Gorucu, S. E., Dang, C., & Shrivastava, V. (2017). A framework for mechanistic modeling of alkali-surfactant-polymer process in an equation-of-state compositional simulator. In *Paper presented at the SPE reservoir simulation conference, Montgomery, Texas, USA, 20–22 Feburary*. https://doi.org/10.2118/182628-MS.

Rivet, S., Lake, L. W., & Pope, G. A. (2010). A coreflood investigation of low-salinity enhanced oil recovery. In *Paper presented at the SPE annual technical conference and exhibition, Florence, Italy, 19–22 September*. https://doi.org/10.2118/134297-MS.

Seccombe, J. C., Lager, A., Webb, K. J., Jerauld, G., & Fueg, E. (2008). Improving wateflood recovery: LoSalTM EOR field evaluation. In *Paper presented at the SPE symposium on improved oil recovery, Tulsa, Oklahoma, USA, 20–23 April*. https://doi.org/10.2118/113480-MS.

Seright, R. S., Fan, T., Wavrik, K., & de Carvalho Balaban, R. (2011). New insights into polymer rheology in porous media. *SPE Journal, 16*(01), 35–42. https://doi.org/10.2118/129200-PA.

Shaddel, S., & Tabatabae-Nejad, S. A. (2015). Alkali/surfactant improved low-salinity waterflooding. *Transport in Porous Media, 106*(3), 621–642. https://doi.org/10.1007/s11242-014-0417-1.

Shaker Shiran, B., & Skauge, A. (2013). Enhanced oil recovery (EOR) by combined low salinity water/polymer flooding. *Energy and Fuels, 27*(3), 1223–1235. https://doi.org/10.1021/ef301538e.

Sheng, J. (2011). *Modern chemical enhanced oil recovery: Theory and practice*. Amsterdam; Boston, MA: Gulf Professional Pub.

Shiran, B. S., & Skauge, A. (2014). Similarities and differences of low salinity polymer and low salinity LPS (linked polymer solutions) for enhanced oil recovery. *Journal of Dispersion Science and Technology, 35*(12), 1656–1664. https://doi.org/10.1080/01932691.2013.879532.

Sorbie, K. S. (1991). *Polymer-improved oil recovery*. Glasgow, Boca Raton, FLA: Blackie: CRC Press.

Suleimanov, B. A., Latifov, Y. A., Veliyev, E. F., & Frampton, H. (2017). Low salinity and low hardness alkali water as displacement agent for secondary and tertiary flooding in sandstones. In *Paper presented at the SPE annual caspian technical conference and exhibition, Baku, Azerbaijan, 1–3 November*. https://doi.org/10.2118/188998-MS.

Suleimanov, B. A., Latifov, Y. A., Veliyev, E. F., & Frampton, H. (2018). Comparative analysis of the EOR mechanisms by using low salinity and low hardness alkaline water. *Journal of Petroleum Science and Engineering, 162*, 35–43. https://doi.org/10.1016/j.petrol.2017.12.005.

Tavassoli, S., Korrani, A. K. N., Pope, G. A., & Sepehrnoori, K. (2016). Low-salinity surfactant flooding—a multimechanistic enhanced-oil-recovery method. *SPE Journal, 21*(03), 744–760. https://doi.org/10.2118/173801-PA.

Teklu, T. W., Li, X., Zhou, Z., Alharthy, N., Wang, L., & Abass, H. (2018). Low-salinity water and surfactants for hydraulic fracturing and EOR of shales. *Journal of Petroleum Science and Engineering, 162*, 367–377. https://doi.org/10.1016/j.petrol.2017.12.057.

Tichelkamp, T., Hosseinzade Khanamiri, H., Nourani, M., Stensen, J.Å., Torsæter, O., & Øye, G. (2016). EOR potential of mixed Alkylbenzene sulfonate surfactant at low salinity and the effect of calcium on "optimal ionic strength". *Energy and Fuels, 30*(4), 2919–2924. https://doi.org/10.1021/acs.energyfuels.6b00282.

Torrijos, P., Iván, D., Puntervold, T., Skule Strand, Austad, T., Bleivik, T. H., et al. (2018). An experimental study of the low salinity smart water – polymer hybrid EOR effect in sandstone material. *Journal of Petroleum Science and Engineering, 164*, 219–229. https://doi.org/10.1016/j.petrol.2018.01.031.

Unsal, E., ten Berge, A. B. G. M., & Wever, D. A. Z. (2018). Low salinity polymer flooding: Lower polymer retention and improved injectivity. *Journal of Petroleum Science and Engineering, 163*, 671–682. https://doi.org/10.1016/j.petrol.2017.10.069.

Vermolen, E. C. M., Pingo Almada, M., Wassing, B. M., Ligthelm, D. J., & Masalmeh, S. K. (2014). Low-salinity polymer flooding: Improving polymer flooding technical feasibility and economics by using low-salinity make-up brine. In *Paper presented at the international petroleum technology conference, Doha, Qatar, 19–22 January*. https://doi.org/10.2523/IPTC-17342-MS.

Wang, J., Ayirala, S. C., AlSofi, A. M., Al-Yousef, A. A., & Aramco, S. (2018). SmartWater synergy with surfactant polymer flooding for efficient oil mobilization in carbonates. In *Paper presented at the SPE EOR conference at oil and gas West Asia, Muscat, Oman, 26–28 March*. https://doi.org/10.2118/190334-MS.

Winsor, P. A. (1948). Hydrotropy, solubilisation and related emulsification processes. *Transactions of the Faraday Society, 44*(0), 376–398. https://doi.org/10.1039/TF9484400376.

Zhang, G., Yu, J., Cheng, Du, & Lee, R. (2015). Formulation of surfactants for very low/high salinity surfactant flooding without alkali. In *Paper presented at the SPE international symposium on oilfield chemistry, The Woodlands, Texas, USA, 13–15 April*. https://doi.org/10.2118/173738-MS.

CHAPTER 5

Hybrid CO_2 EOR Using Low-Salinity and Smart Waterflood

ABSTRACT

The CO_2 injection is the widely used enhanced oil recovery (EOR) method because of economics and recovery efficiency. Although miscible/immiscible CO_2 injections mainly enhance the displacement efficiency employing a number of advantages, the CO_2 injection has an inherent unfavorable mobility ratio because of low-viscous CO_2 gas. To support the EOR potential of CO_2 injection, auxiliary water injection is developed to cooperate in CO_2 injection. The coapplication of water and CO_2 injections enhances displacement efficiency as well as sweep efficiency improving the mobility ratio. Recently, the use of low-salinity water rather than conventional high-salinity water in coinjections of water and CO_2 is proposed as low salinity–augmented CO_2 EOR (LS-CO_2 EOR). This chapter reviews the studies of CO_2 EOR, mainly CO_2 water-alternating gas and carbonated water injection, to investigate the effect of salinity on the mechanisms of CO_2 EOR process. Laboratory and numerical simulation studies of LS-CO_2 EOR are discussed to evaluate whether the additional displacement efficiency employing the mechanism of low-salinity waterflood or other effects can be introduced.

The understanding of CO_2 EOR has been investigated, and some researchers have reported the experimental works of carbonated water recovery as early as 1951 (Martin, 1951). The first patent of CO_2 EOR technology was granted by the Atlantic Refinery Company in 1952 (Whorton, Brownscombe, & Dyes, 1952). In 1964, the field test of CO_2 EOR was developed in the Mead Strawn Field and the test was designed with carbonated water injection following CO_2 slug (Holm, apos, & Brien, 1971; Holm, 1987). The CO_2 EOR produces more oil by 53%–82% over the best performance of waterflood. Following the success of pilot test, the first commercial CO_2 EOR project was deployed in SACROC (Scurry Area Canyon Reef Operators Committee) Unit of the Kelly-Snyder Field in Scurry County, West Texas (Langston, Hoadley, & Young, 1988). The successful field project leads to a number of applications of commercial CO_2 EOR projects. The CO_2 EOR can be applied with various designs (Jarrell, Fox, Stein, & Webb, 2002): (1) continuous CO_2 gas injection (CGI), (2) continuous CO_2 injection followed with water, (3) conventional water-alternating gas (WAG) followed with water, (4) tapered WAG, and (5) WAG followed gas. Using the water, the CO_2 EOR can improve the sweep efficiency as well as displacement efficiency. Recently, the usage of low-salinity water or smart water rather than conventional water into the CO_2 EOR has been proposed to introduce the synergy. The applications of hybrid low salinity–augmented CO_2 EOR (LS-CO_2 EOR) are discussed in this chapter. Before the discussion of recent investigations of the LS-CO_2 EOR, the important aspects of CO_2 injection are firstly summarized to help the understanding of LS-CO_2 EOR. The detailed description of CO_2 flood will be found in the references (Jarrell et al., 2002; Stalkup, 1983).

MINIMUM MISCIBLE PRESSURE

Although there are other solvents, the CO_2 is a good solvent for miscible process. The CO_2 generally forms a phase, which has comparable density of a liquid and low viscosity, in a high pressure condition. More importantly, the CO_2 relatively has a low minimum miscible pressure (MMP), which is defined as minimum pressure to achieve a miscibility, compared with the methane and nitrogen. The miscible process can be effective for a reservoir system, which has a higher pressure than MMP. Because the CO_2 generally results in a lower MMP compared with the other solvents, the CO_2 injection can be used to enhance the miscible oil recovery process in a wide range of oil reservoirs.

MISCIBLE/IMMISCIBLE PROCESSES

The solvent EOR process can be either a miscible or an immiscible displacement depending on the miscibility condition. The miscibility between two fluids, i.e., displacing and displaced fluids, can be achieved by first-contact miscibility or multiple-contact miscibility.

Hybrid Enhanced Oil Recovery using Smart Waterflooding. https://doi.org/10.1016/B978-0-12-816776-2.00005-2

Because the first-contact miscibility between the displacing and displaced fluids results in a completely single phase, the displacing fluid is able to recover all displaced fluid contacted. The miscible displacement achieving the first-contact miscibility indicates that the injecting gas is miscible in all proportions with the reservoir fluid under the prevailing reservoir temperature and pressure conditions. It is termed as first-contact miscible process.

Another immiscible displacement can be classified with two processes (Slobod & Koch, 1953), and it corresponds to the CO_2 injection in the real field. In the first immiscible process of type I, the equilibrium phases at the gas-oil front area are essentially immiscible. In other words, the CO_2-rich phase displaces oil-rich phase with the interface between the two phases. It introduces the advantages of more viscous displacing gas, oil swelling, and oil viscosity reduction. In the second immiscible flood of type M, the injected gas is sufficiently enriched at the front as to be completely miscible with the reservoir fluid. It is the miscible process achieving multiple-contact miscibility. In the CO_2 injection, CO_2- and oil-rich phases are miscible on not the first contact but multiple contacts. During the multiple contacts between oil- and CO_2-rich phases, the mass transfer between oil and CO_2 proceeds until the oil-rich phase cannot be distinguished from the CO_2-rich phase. The transfer process between oil- and CO_2-rich phases is described with condensing/vaporizing mechanisms (Zick, 1986). Following the condensing mechanism, the CO_2 firstly condenses into the hydrocarbon liquid, i.e., oil-rich phase. The condensation makes the oil-rich phase lighter and, often, drives some lightest methane out ahead of oil bank. Because of the succeeding vaporizing mechanism, the lighter oil components vaporize into the CO_2-rich phase. The vaporization makes the CO_2-rich phase denser to be soluble in the oil phase. When the multiple-contact miscibility between the CO_2- and oil-rich phases is achieved, the two phases cannot be distinguishable in terms of properties. During the multiple-contact miscible process of CO_2 injection, there are some advantages of viscosity ratio, oil swelling, and condensing/vaporization to recover oil. The miscibility between CO_2 and oil is sensitive to the reservoir fluid composition, pressure, and temperature. Because the real production commonly undergoes little change in the temperature except for the thermal EOR process, the main interesting factor on the miscibility is the pressure. The condensing CO_2 in oil-rich phase and vaporizing oil in CO_2-rich phase increase with an increase in pressure.

DETERMINATION OF SOLVENT PHASE BEHAVIOR

There are several experiments to characterize the solvent phase behavior such as single contact, multiple contact, and slim tube tests (Lake, 1989). The single contact experiment is useful to construct the *P-z* diagram, which only shows the number and types of phases and liquid volume, since the pressure can be changed. The multiple contact experiment can formulate the ternary diagram and imparts the compositional information, which is not described from *P-z* diagram. The experiment of slim tube displacement can bridge the static experiment and coreflood. Although it does not exactly mimic the displacement process, it still offers the dynamic properties of phase behavior. Although the slim tube test measures an effluent concentration, its main objective is to identify the MMP. The pressure, in which certain amount of oil recovery is produced, is the MMP or minimum dynamic miscibility pressure. Although there are a number of definitions to indicate MMP, they show the similar trends in correlations. Corresponding to MMP, the critical tie line passes through the crude composition. The MMP is conventionally less than the pressure of first-contact miscibility. The determination of MMP using slim tube test requires a high cost. Therefore, alternative approaches to determine MMP are (1) mathematical models and (2) thermodynamic MMP correlations (Jarrell et al., 2002). The mathematical model uses phase equilibrium data and EOS to estimate the thermodynamic MMP. Once the appropriate data are given, an excellent determination of MMP is possible at low cost. When the acquisition of phase equilibrium data is unavailable, MMP correlations can be applied for the specific condition.

SOLUBILITY OF CO_2 IN WATER

The solubility of most hydrocarbon components in water is negligible over the range of temperature and pressure conditions. However, the solubility of CO_2 in water is much higher than that of hydrocarbon components in water. The solubility of CO_2 is a function of pressure, temperature, and salinity. Søreide and Whitson (1992) developed the numerical model of PR EOS considering the temperature-, pressure-, and salinity-dependent aqueous solubility of CO_2 based on the experimental database. Chang, Coats, and Nolen (1998) also developed the EOS model using CO_2 solubility, as well as the properties of CO_2-saturated water including formation volume factor, compressibility, and viscosity based on the experimental database.

EFFECT OF SALINITY ON THE SOLUBILITY IN WATER

The presence of salt or electrolyte in an aqueous solution modifies the solubility of organic components in water. The phenomena of salting-out effect have been termed less soluble organic components in aqueous salt solutions than in pure water. The empirical Setschenow equation (Setschenow, 1889) has been developed to describe this effect. Until the work of Harvey and Prausnitz (1989), the investigation of EOS model considering the salinity-dependent solubility is limited. Most of the studies considering the effect are applicable to only low pressure and low temperature conditions. In addition, they used the methods that are not appropriate to the system containing supercritical gases. Harvey and Prausnitz (1989) developed the EOS model to be applicable to the high pressure condition. The developed EOS model accounts for the intermolecular effects by the ionic effects and conventional intermolecular forces for nonelectrolytes. The model also uses the Setschenow constant. Although the model accurately predicts the salting-out effects under the low salt condition less than 1 molal concentration, it underpredicts the salting-out effect in moderate to high salt conditions. Søreide and Whitson (1992) provided the novel approach to predict the mutual solubilities of brine/hydrocarbon mixtures with an EOS at high pressure and high temperature conditions in addition to describing the effect of salinity in aqueous phase. The model accurately predicts the experimental results of hydrocarbons, CO_2, N_2, and H_2S in pure water and NaCl brine.

CO_2 WATER-ALTERNATING GAS INJECTION

The injection of solvents has been of interest to the additional oil extraction from porous media. The solvent injection recovers the oil by incorporating the extraction, dissolution, vaporization, solubilization, condensation, or other phase behavior changes with crude oil, which contribute the oil recovery mechanisms including the oil viscosity reduction, oil swelling, solution gas drive, and oil extraction. In the early 1960s, the small injection of liquefied petroleum gas (LPG) has been investigated for the solvent process. The high cost of the LPG is a barrier for the practical implementation. In the late 1970s, the CO_2 has become the widely used solvent. Orr and Taber (1984) proposed the usage of CO_2 for EOR process. The CO_2 injection has a number of injection process designs, which include the combination of continuous, alternating, and chase fluid injection schemes. The miscible gas-water injection, originally proposed by Caudle and Dyes (1958), uses the water to support the CO_2 EOR process. The first interesting scheme of miscible gas-water injection is the simultaneous water and CO_2 injection (Caudle & Dyes, 1958). Then, the CO_2 water-alternating gas (CO_2 WAG) process is proposed and extensively investigated (Christensen, Stenby, & Skauge, 2001; Rogers & Grigg, 2001). The water injection of the CO_2 WAG process offers an improvement in an inherent unfavorable mobility ratio between the solvent and oil to secure the recovery mechanism of CO_2 EOR process. Some studies of CO_2 WAG have explored the effects of CO_2 solubility into water and brine salinity on the performance of CO_2 WAG process. When the injected CO_2 contacts water during CO_2 WAG process, some of CO_2 dissolves in water. The CO_2 dissolution in water might prevent some of CO_2 from contacting oil, contribute to lower oil recovery efficiency, and change the formation volume and viscosity of water. The CO_2 solubility in brine is sensitive to pressure, temperature, and salinity of the water. The CO_2 solubility in brine increases with the increasing pressure and decreasing salinity. In addition, the recent studies of CO_2 WAG process have been interested in the usage of low-salinity water or smart water, rather than conventional high-salinity water, during the water injection period. The studies have investigated whether the low salinity–augmented CO_2 WAG (LS-CO_2 WAG) process introduces both mechanisms of CO_2 injection and LSWF.

EXPERIMENTS

Kulkarni and Rao (2005) investigated the performances of miscible and immiscible CO_2 WAG processes comparing with CGI and suggested the optimum design of CO_2 WAG process to maximize economics. In the analysis of CO_2 WAG, the effect of brine salinity is investigated by using both 5% NaCl brine and multicomponent reservoir brine from the Yates Field in West Texas. The n-decane and Berea sandstone core are subjected to the experiments. Before the coreflooding experiments, a couple of approaches determine the MMP of decane to define the displacement experiments to be either miscible or immiscible condition. Based on the data from literature studies, empirical equation, and numerical simulation, the MMP of the decane is estimated as 1880 psi. Immiscible and miscible floods are performed at 500 and 2500 psi, respectively. In both conditions, the secondary recovery is performed with waterflood. The tertiary CGI and CO_2 WAG follow, respectively. During the water injection, the two brines of 5% NaCl and Yates Field brine of 9200 mg/L TDS are used. During the secondary waterflood using both high- and low-salinity brines, the cores used in the experiments are determined as water-wet. The performances of CGI and CO_2 WAG processes are assessed with CO_2 utilization factor and tertiary recovery factor. The CO_2 utilization factor is defined as the volume of CO_2 gas injected under the standard condition per a barrel of oil as defined in Eq. (5.1). It is conveniently used to evaluate the feasibility of CO_2 injection project. Tertiary recovery factor, defined as Eq. (5.2), is used to normalize the tertiary oil production of CO_2 injection after secondary waterflood. In contrast to the simple oil recovery analysis, they are more appropriate to determine the contribution of CO_2 injected on the oil production by considering both the CO_2 amount injected and remained oil amount.

$$UF_{CO_2} = \frac{V_{CO_2\ injection}}{V_{oil\ production}} \tag{5.1}$$

$$TRF = \frac{V_{oil\ production}}{V_{residual\ oil,\ waterflood} V_{total\ CO_2\ injection}} \tag{5.2}$$

where UF_{CO_2} indicates the CO_2 utilization factor; $V_{CO_2\ injection}$ is the total volume of CO_2 injected; $V_{oil\ production}$ is the total volume of oil produced; TRF is the tertiary recovery factor; $V_{residual\ oil,\ waterflood}$ is the residual oil volume after waterflood; and $V_{total\ CO_2\ injection}$ is the total pore volume of CO_2 injection.

When the 5% NaCl brine is injected, tertiary CGI shows higher oil recovery by 9.2% than tertiary CO_2 WAG in the miscible condition (Fig. 5.1A). In the immiscible condition, both CGI and CO_2 WAG have similar oil recovery of 23% (Fig. 5.1A). The higher oil recovery of CGI is attributed to the higher CO_2 amount injected. The analysis of tertiary recovery factor reasonably compares the performances of CGI and CO_2 WAG processes (Fig. 5.1B). The higher TRF is observed in CO_2 WAG over CGI in both miscible and immiscible conditions. The lower TRF of CGI implies the lower economics compared with CO_2 WAG process. Interestingly, the oil recovery by CO_2 injection in both miscible and immiscible conditions is delayed (Fig. 5.1B). The other CGI and CO_2 WAG processes using Yates Field brine draw similar trend of total oil recovery and tertiary recovery factor (Fig. 5.2). Miscible condition is more favorable to oil recovery in both processes than immiscible condition. Although CGI shows higher oil production than CO_2 WAG, tertiary recovery factor shows opposite result. Regardless of brine types, the CO_2 WAG is determined to produce higher economics of tertiary recovery than CGI, especially in miscible condition. Because the delay in oil production is also observed in immiscible condition, not in miscible condition, another assessment is performed to investigate the factors contributing the oil production delay. The delay in oil breakthrough is hardly observed in miscible condition because of an increasing CO_2 density. In the immiscible CGI process, brine type changes and the delay in oil production is evaluated. The higher salinity and monovalent condition injecting 5% NaCl brine shows less delay in oil production compared with the low salinity and multicomponent condition injecting the Yates Field brine. The degree of delay in oil production is explained with the CO_2 solubility in immiscible condition. Additional coreflooding confirms the role of CO_2 solubility in brine on the delay in oil production. After the secondary waterflood using CO_2-saturated Yates Field brine or using normal Yates Field brine, the tertiary recovery of CGI is investigated. In the results of tertiary recovery factor, the secondary injection of CO_2-saturated brine hastens the oil production of tertiary CGI. The observation from the coreflooding confirms the role of salinity- or ionic composition-dependent CO_2 solubility in brine on the oil production of CO_2 injection. This experimental study might not demonstrate the synergetic effects between LSWF and CO_2 injection. However, it is obvious that the ionic composition and salinity of brine can be crucial factors to influence the performance of CO_2 injection.

Aleidan and Mamora (2010) experimentally investigated the effects of lowering water salinity on the oil recovery of coreflooding of simultaneous water and CO_2 injection and CO_2 WAG processes. The study also carried out the waterflood and continuous CO_2 injection processes for a quantitative comparison. Before the limestone coreflooding experiments, the slim tube experiment measures the MMP of West Texas dead oil with 31°API. The MMP is determined as 1800 psi at which the oil recovery reaches to 90%. Making the miscible condition system, the coreflooding experiments are conducted at 1900 psi and 120°F. To observe the effect of salinity, the salinity of brines is adjusted by controlling the NaCl concentration. For the waterflood experiment, two salinity levels of 0 and 6 wt% are used. For the WAG process experiment, three salinity levels of 0, 6, and 20 wt% are prepared. The waterflood experiments injecting the distilled water and saline water show the equivalent oil recovery of 54% produced. Lowering salinity in brine hardly contributes to increase the oil production. It is explained that the system of waterflood does not satisfy the suggested conditions to provoke the mechanism of LSWF in carbonate rocks. The brine does not have any potential-determining ions, and the temperature is low. These results screen

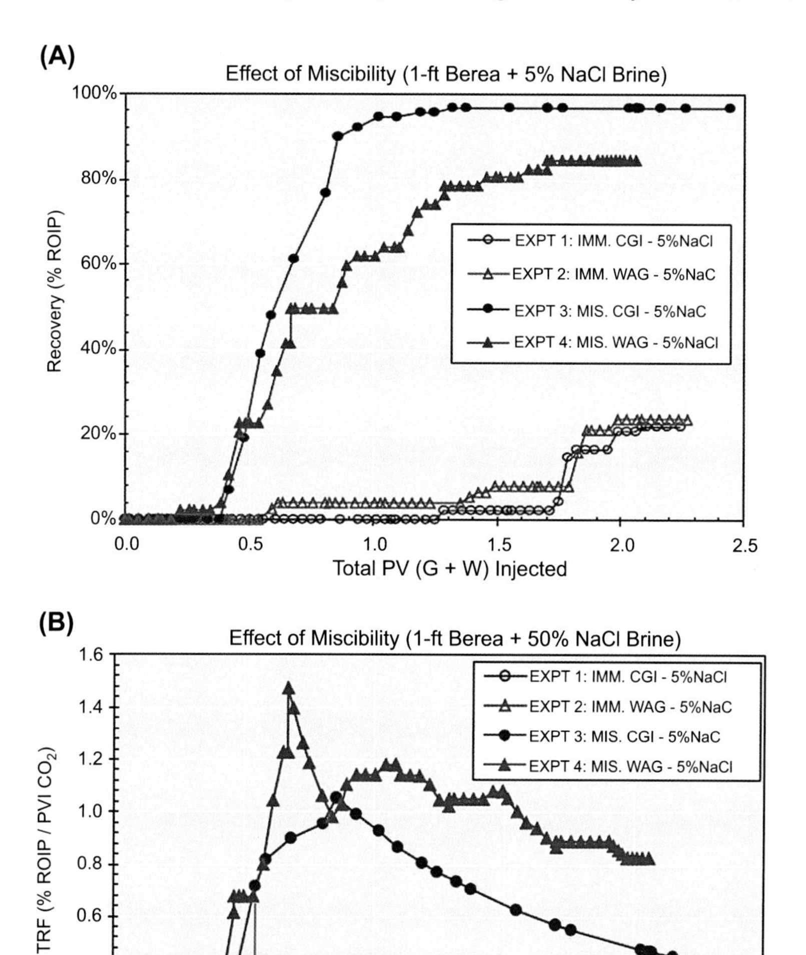

FIG. 5.1 The **(A)** oil recovery and **(B)** tertiary recovery factor of continuous CO_2 gas injection and CO_2 water-alternating gas process using 5% NaCl brine in miscible and immiscible conditions. (Credit: From Kulkarni, M. M., & Rao, D. N. (2005). Experimental investigation of miscible and immiscible water-alternating-gas (WAG) process performance. *Journal of Petroleum Science and Engineering, 48*(1), 120. https://doi.org/10.1016/j.petrol.2005.05.001.)

out the potential of LSWF mechanism during the coinjection and alternating injection of CO_2 and low-salinity water. Therefore, any contribution by using low-salinity water on oil recovery is attributed to the CO_2 solubility in water. The coreflooding of CGI produces the additional oil by 21% over waterflood because of miscibility between CO_2 and oil.

The CO_2 WAG process produces the oil recovery in the range of 74.3%–92.7% depending on the water salinity. As the injected water has lower salinity, higher oil recovery is measured. As the brines have salinities of 6 and 20 wt%, they are estimated to have lower solubilities of CO_2 in water by at least 25% and 60% compared with the distilled water. The relationship between salinity and CO_2 solubility is also confirmed through the CO_2 production at the outlet. Another coreflooding of the simultaneous water and CO_2 injection is investigated by controlling the salinity

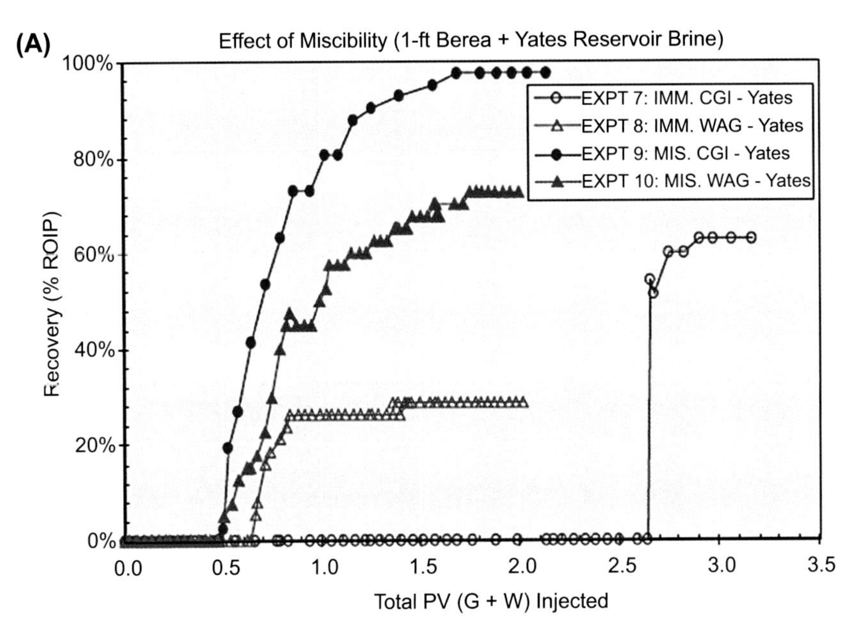

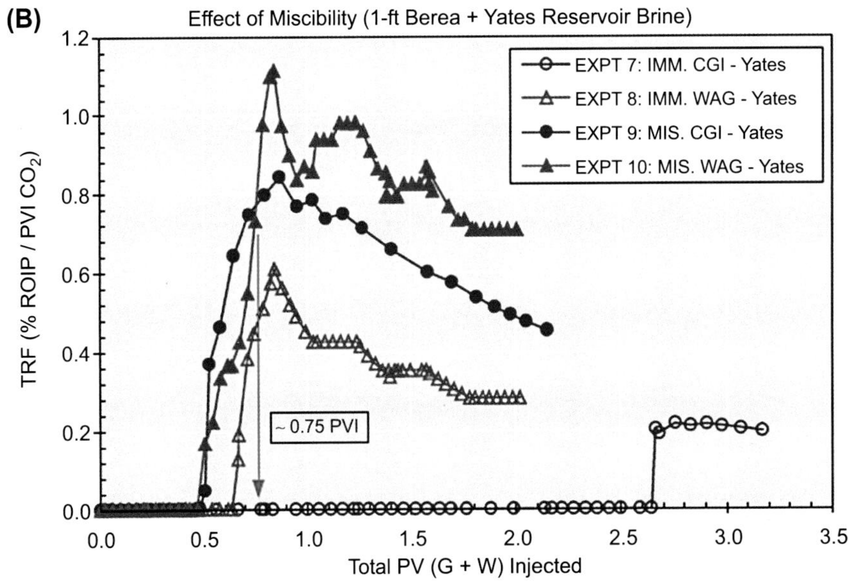

FIG. 5.2 The **(A)** oil recovery and **(B)** tertiary recovery factor of continuous CO_2 gas injection and CO_2 water-alternating gas process using Yates Field brine in miscible and immiscible conditions. (Credit: From Kulkarni, M. M., & Rao, D. N. (2005). Experimental investigation of miscible and immiscible water-alternating-gas (WAG) process performance. *Journal of Petroleum Science and Engineering, 48*(1), 120. https://doi.org/10.1016/j.petrol.2005.05.001.)

of brine. Depending on the salinity of brine, the oil recoveries from 81.5% to 98.6% are observed. Similar trends of oil recovery and CO_2 solubility in water depending on the salinity are also observed. Previously, the increasing solubility of CO_2 in water has been considered as the loss of CO_2, which is supposed to contact oil. However, the experimental observations of this study indicate the positive aspects of CO_2 solubility in water on the oil recovery and less CO_2 requirement injected in low salinity condition. It is concluded that the mixture of CO_2-water following CO_2 slug at the displacement front is effective to contact the bypassed oil by the CO_2 slug. The mixture of CO_2-water improves the displacement efficiency and increases the oil recovery of the injection process of CO_2 and water.

Jiang, Nuryaningsih, and Adidharma (2010) analyzed the effect of salinity as well as divalent cation on the coreflooding of miscible CO_2 WAG process using synthetic oil, which is composed of n-decane and n-hexadecane, and crude oil. Two different types of connate brines are prepared: NaCl brine of 1000 ppm, mixing brine of NaCl of 4000 ppm, and $CaCl_2$ of 4000 ppm. The synthetic and crude oils have MMPs as 1673 and 2541 psi at 60°C, respectively. In the coreflooding experiment, the CO_2 WAG process is applied in the tertiary mode after the secondary waterflooding. The miscible process during CO_2 WAG is achieved by making system to be at 20% higher pressure than the MMP. During the CO_2 WAG, the slug sizes of both CO_2 and water are 0.25 PV and the volumetric WAG ratio is set as 1:1.

The first set of coreflooding using the synthetic oil quantifies the effects of salinity and divalent cations on the oil recovery of secondary waterflood, tertiary CO_2 WAG, and chasing waterflood. The core is initially saturated with the NaCl brine of 1000 ppm TDS. In the secondary waterflood, the increasing salinity of injecting brine from 1000 to 32,000 ppm TDS results in the negligible change in oil production and 32% of oil recovery is observed. The negligible effect of salinity indicates that the mechanism of LSWF hardly exists. During the period of tertiary CO_2 WAG, the slight increase of oil recovery by 5.6% is observed with an increasing salinity. The observation is in contrast to the experimental observations of Aleidan and Mamora (2010). The improved oil recovery in higher salinity condition explains that the salting-out effect, which indicates the less solubility with an increase in salinity, makes more CO_2 to be miscible with oil. In another coreflooding experiment, the effect of divalent cation is investigated and the negligible effect of divalent cation is observed on the oil recovery. In the second set of coreflooding using crude oil, the role of salinity on the oil recovery of secondary waterflood and tertiary CO_2 WAG is investigated. The core is saturated with connate brine of NaCl of 20,000 ppm and $CaCl_2$ of 10,000 ppm. In contrast to the first set coreflooding using synthetic oil, the oil recovery of waterflood increases with a decrease in salinity. The trend of waterflood recovery depending on the salinity can be attributed to the effect of LSWF. However, the lower-salinity water injection of CO_2 WAG shows the slightly less reduction in tertiary recovery. Although the unfavorable effect of low-salinity water on the performance of tertiary CO_2 WAG process is observed, the application of secondary LSWF can enhance the oil recovery ahead of tertiary CO_2 WAG.

A couple of studies (Teklu et al., 2014, 2016) have investigated the combination of miscible and immiscible CO_2 injection and low-salinity waterflood. The displacement experiments are designed with one cycle CO_2 WAG process after secondary waterflood. In the CO_2 WAG process, CO_2 injection follows the LSWF. Before the coreflooding experiment, MMP, IFT, and contact angle are measured under various conditions. Brines of connate water, seawater, three cases of synthetic low-salinity water, and deionized water are prepared in the descending salinity order. Experimentally, the rising bubble apparatus (RBA) tool estimates the MMP of crude oil and CO_2 as 2500 psi. The numerical approach using multiple mixing cell (MMC) calculates the MMP as 2470 psi being comparable with the experimental determination. Pendant drop method determines the IFTs of crude oil and the various brines. As the salinity decreases, the IFT slightly increases. The experiment of captive oil-bubble contact angle measures the contact angle in the various systems of crude oil/brine/rock. Three rocks of different mineralogy are prepared, including carbonate, Berea sandstone, and Three Forks. Preliminary, the effect of initial wetness is investigated by using aged cores or unaged cores. Regardless of the aging status of cores, the consistent tendency is observed of the decreasing contact angle with a decrease in salinity. Additional measurement of contact angle using the aged cores analyzes the contact angle in the system of crude oil/mixture fluid of CO_2 and brine/rock at miscible condition of 2500 psi (Fig. 5.3). It is clearly observed that the contact angle is decreasing with a decrease in salinity. The additional IFT measurement in the system of crude oil/mixture fluid of CO_2 and brine also results in the slight reduction in IFT by adding CO_2 at atmospheric condition. The results of IFT and contact angle measurements imply that the wettability can be improved by the usage of low-salinity water and addition of CO_2. However, the IFT reduction is only effective by the addition of CO_2. LS-CO_2 WAG is determined to introduce both wettability modification and miscible effects.

The EOR potential by the LS-CO_2 WAG is evaluated through the coreflooding in miscible and immiscible conditions. The two sets of coreflooding using carbonate cores are carried out in the miscible condition. The preflush of LSWF produces the additional oil recovery

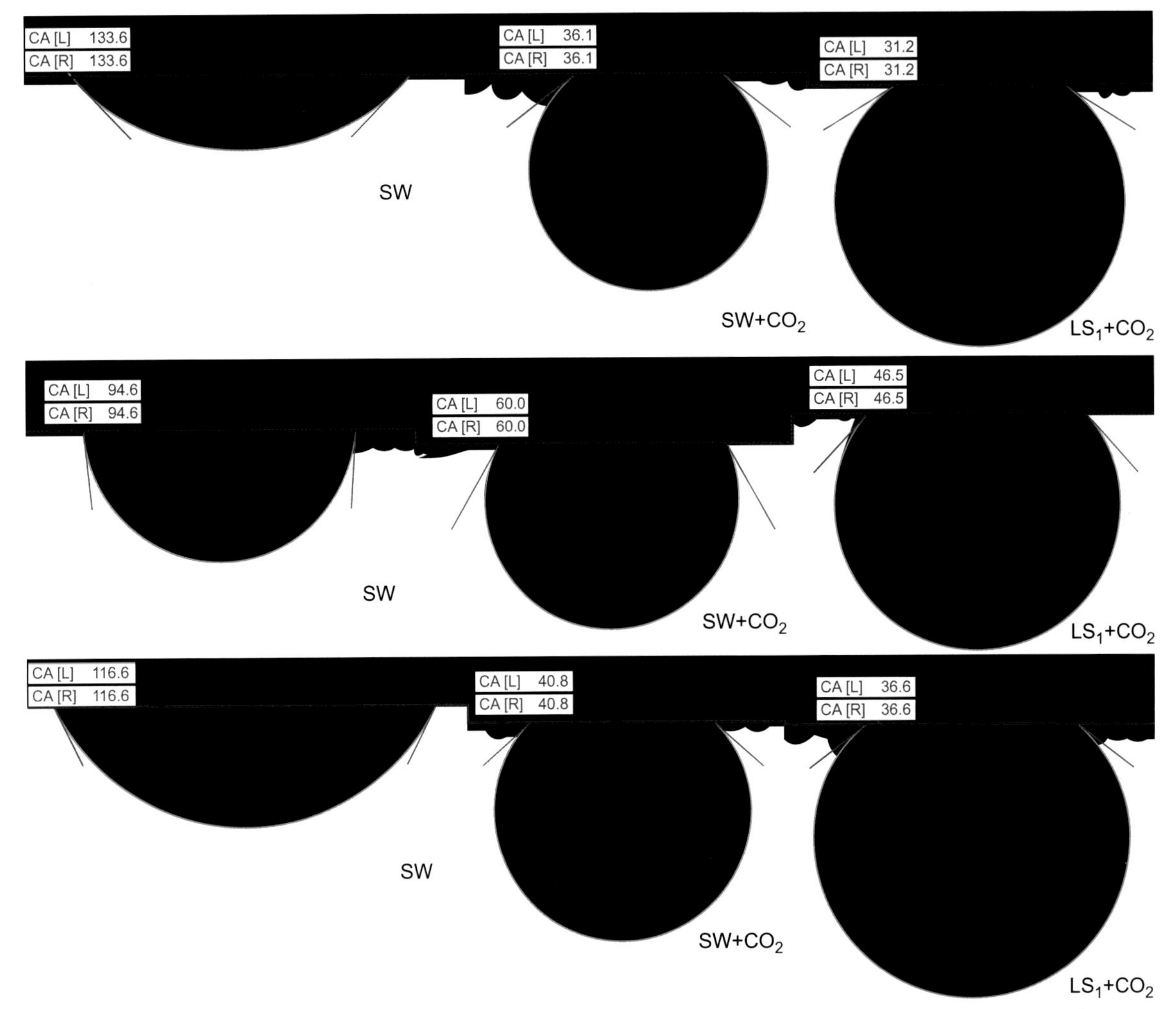

FIG. 5.3 The contact angle of the system of crude oil/mixture fluid of brine and CO_2/rock varying brine and rock types. (Credit: From Teklu, T. W., Alameri, W., Graves, R. M., Kazemi, H., & AlSumaiti, A. M. (2016). Low-salinity water-alternating-CO_2 EOR. *Journal of Petroleum Science and Engineering, 142*, 101–118. https://doi.org/10.1016/j.petrol.2016.01.031.)

by 7.1% and 5.6% over the secondary waterflood recovery. The miscible CO_2 injection process enhances the oil recovery by 14.2% and 25% over the recovery of the LSWF, respectively. The third set of coreflooding using sandstone core is performed at immiscible condition. The LSWF increases the oil recovery by 7.7% over the secondary waterflood. The CO_2 injection significantly produces more oil by 38.5% than LSWF. In conclusion, the LS-CO_2 WAG is effective to improve the oil recovery by IFT reduction and wettability modification. The additions of CO_2 and low-salinity water are effective to change both the carbonates and sandstones toward hydrophilic conditions. The addition of CO_2 has another advantage to reduce the IFT between oil and brine. The experimental study obviously demonstrated that the LS-CO_2 WAG process can secure the both mechanisms of LSWF and miscible/immiscible CO_2 injection.

Ramanathan, Shehata, and Nasr-El-Din (2015) quantified the effect of salinity-dependent CO_2 solubility on the immiscible CO_2 WAG process through the contact angle measurements and coreflooding experiments. The captive bubble method measures the contact angle in the two different systems of crude oil/Berea sandstone/brine and crude oil/Berea sandstone/brine/CO_2. The aged and unaged cores are investigated to analyze the effect of initial wetness. Three different brines include the formation brine of 174,156 ppm TDS, seawater of 54,680 ppm TDS, and NaCl brine of 5000 ppm TDS as low-salinity water. In the system of crude oil/sandstone/brine, the low-salinity water shows the lowest contact angle and seawater shows the highest contact angle for both aged and unaged cores. During the measurement, the temperature of the system is increased to examine the effect of temperature on the contact angle. The higher contact angle with an increase in temperature is observed. Interestingly, the formation water condition has lower contact angle than the seawater despite higher salinity, i.e., the less water-wetness in formation water condition than seawater condition. It is explained that the seawater, which has the relatively higher fraction of divalent cations compared with formation water, results in the bridging between the anionic surfactant, i.e., negative polar component, of crude oil and the significant amount of divalent cations on the rock surface. Another experiment measures a dynamic contact angle in the system of crude oil/brine/CO_2/unaged rock for 24 h. Because of the interactions between oil and CO_2, and oil and brine, the measured contact angle has some fluctuations. After some time to be stabilized, the consistent tendency between contact angle and brine type is similar to the previous measurements. The results of contact angle measurements indicate that low-salinity water results in the more water-wet rock samples. The seawater brine shows the more oil-wet rock samples in both systems representing waterflood and CO_2 WAG process.

In a number of coreflooding experiments, the secondary CO_2 WAG processes and waterflood using low-salinity water and seawater are analyzed. The experiments also use both the aged and unaged cores to investigate the effect of initial wetness. Firstly, the conventional waterflood using seawater and LSWF are investigated, considering the initial wetness of cores. In the system of unaged core, the effect of LSWF is hardly observed. Instead, a slight higher oil recovery is observed in conventional waterflood compared with the LSWF. In the system of aged core, which is initially more oil-wet condition, the higher oil recovery by 14% is observed for the LSWF than the conventional waterflood. In addition, the LSWF produces the significantly higher oil recovery in aged core system than unaged core system. However, the conventional waterflood shows the insignificant change of oil recovery between both systems. These observations are in line with the results of contact angle measurement. It is obvious that the LSWF is effective to modify the wettability of oil-wet rock system. The measured pressure drop also implies no formation damage of fines migration, precipitation, or plugging. Secondly, the immiscible CO_2 WAG processes using seawater and low-salinity water are investigated (Fig. 5.4). The coreflooding of CO_2 WAG process uses only unaged core, which screens out the wettability modification effect of LSWF. In the comparison between conventional CO_2 WAG using seawater and LS-CO_2 WAG processes, conventional CO_2 WAG shows the slightly higher oil recovery over the LS-CO_2 WAG. It is explained that the slight increment by 2.93% is attributed to the salting-out effect, which is similar to the observations of Jiang et al. (2010). The higher solubility of CO_2 in low-salinity water results in the less contact between oil and CO_2. The less contact leads to the less contribution of immiscible mechanism of CO_2 injection on the oil production. The fluctuating pressure drop is observed during the period of CO_2 injection for both conventional CO_2 WAG and LS-CO_2 WAG processes. It is estimated to the result of CO_2 dissolution in water. This study clearly confirmed the effect of salinity-dependent CO_2 solubility on the performance of LS-CO_2 WAG process in water-wet system. Because this study lacks the investigation of LS-CO_2 WAG process using oil-wet cores, it is not clarified whether LS-CO_2 WAG process secures the mechanism of LSWF or not.

Ramanathan, Shehata, and Nasr-El-Din (2016) reported the further investigations of CO_2 WAG process with various brine types. The axisymmetric drop shape analysis measures the IFT of the brine/crude oil/N_2 and brine/crude oil/CO_2 systems. The same brines of formation brine, seawater, and low-salinity water are prepared. It is clearly measured that high-salinity brine shows the higher density, and higher temperature decreases the density of brine. In the system of brine/crude oil/N_2, the equilibrium IFT is measured at various temperature and immiscible conditions. The low-salinity water shows the highest IFT and seawater results in lowest IFT regardless of temperature condition. Because the effect of monovalent ions on interfacial activity between oil and water is weak and interfacial active substances are oil-soluble, it is not able to influence the interfacial interactions (Bai, Fan, Nan, Li, & Bao-Shi, 2010). The effect of temperature on the IFT shows a similar trend of increasing IFT with an increase in temperature. Referring the experiments of Hjelmeland and Larrondo (1986), there will be a lower concentration of the surface-active components at the brine/oil interface at higher temperatures. As a result, higher temperature condition might contribute to the higher IFT. The dynamic interfacial tension is measured in the

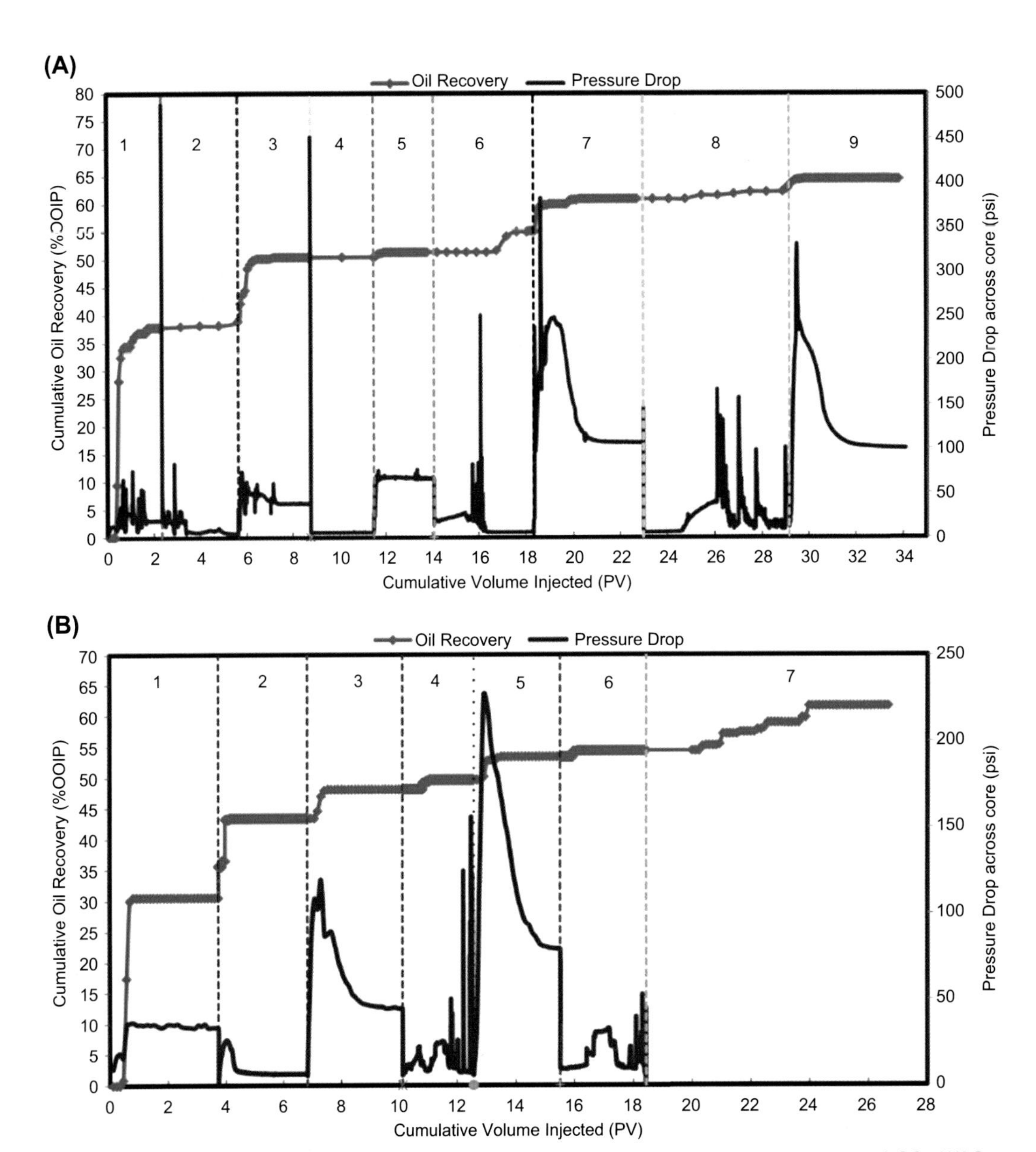

FIG. 5.4 The oil recovery and pressure drop across the unaged core for the experiments of CO_2 WAG process using **(A)** seawater and **(B)** low-salinity water. (Credit: From Ramanathan, R., Shehata, A. M., & Nasr-El-Din, H. A. (2015). Water alternating CO_2 injection process – does modifying the salinity of injected brine improve oil recovery? *Paper presented at the OTC Brasil, Rio de Janeiro, Brazil, 27–29 October*. https://doi.org/10.4043/26253-MS.)

system of the brine/oil/CO_2 by changing the brine type. Because of the interaction of CO_2 in both brine and oil, a slight increase in IFT is observed at initial period. When there is no effective mass transfer across the phases, the stabilized IFT and the consistent tendency of IFT depending on the salinity are obtained. The additional dynamic contact angle measurement of CO_2 WAG using aged Berea sandstone rock is supplemented. In the system using aged rock, the low-salinity water shows the lowest contact angle compared with the formation water and seawater. The seawater shows the highest contact angle. The observations indicate the potential of LS-CO_2 WAG to modify wettability toward more water-wet condition.

Two coreflooding of conventional CO_2 WAG and LS-CO_2 WAG is performed using aged core (Fig. 5.5). Although conventional CO_2 WAG recovers oil up to 76.1%, the LS-CO_2 WAG recovers by 97.7%. Recalling the previous experimental observation of CO_2 WAG processes using unaged core (Ramanathan et al., 2015), the opposite observation is reported in the study (Ramanathan et al., 2016). The performance of LS-CO_2 WAG process obviously prevails the performance of conventional CO_2 WAG process in aged core system. The 36% more oil is recovered despite salting-out effect. Higher oil recovery of LS-CO_2 WAG process than conventional CO_2 WAG process corresponds to contact angle measurement. In the aged core system, the usage of low-salinity water is favorable to the both waterflood (Ramanathan et al., 2015) and CO_2 WAG process (Ramanathan et al., 2016). Lastly, the profiles of saturation and porosity in the four cores, which are used in the coreflooding experiments of conventional CO_2 WAG process and LS-CO_2 WAG process, are estimated by CT scanning method. Two cores are aged cores used in the work of Ramanathan et al. (2016) and last two cores are unaged cores used in the work of Ramanathan et al. (2015). Different profiles in the saturation distribution and average porosity are clearly observed. Both studies have demonstrated the synergetic potential of immiscible LS-CO_2 WAG process through comprehensive experiments. It is concluded that the LS-CO_2 WAG is able to introduce both effects of LSWF and CO_2 injection depending on the initial wetness. Although the low-salinity water is unfavorable to reduce IFT, it is effective to modify the wettability of less water-wet core toward more water-wet. Although the salting-out effect is slightly unfavorable to the contacting between oil and CO_2, the effect of immiscible CO_2 injection sufficiently enhances oil production. Therefore, the LS-CO_2 WAG process is a promising hybrid EOR process securing synergy.

Kumar, Shehata, and Nasr-El-Din (2016) reported the coreflooding experiments of LSWF and LS-CO_2 WAG processes. The experiments of LSWF examine the promising ionic composition of brine to introduce the mechanism of LSWF into LS-CO_2 WAG process. The brines of formation brine, seawater, and three types of low-salinity water are prepared. The low-salinity waters are NaCl brine, KCl brine, and $MgCl_2$ brine, and they equally have 5000 ppm TDS. In the experiments of secondary waterflood using the brines, the higher recovery is observed in LSWF injecting NaCl brine. Two coreflooding experiments injecting CO_2 and NaCl brine are investigated in immiscible condition. The injection processes are designed to apply the secondary LSWF using the NaCl brine and tertiary CO_2 injection, and secondary LS-CO_2 WAG using NaCl brine. In the first coreflooding, the secondary LSWF using NaCl brine recovers 36.32% of oil and tertiary immiscible CO_2 injection increases the recovery by 7.7%. The second coreflooding of LS-CO_2 WAG using NaCl brine produces oil recovery up to 66.84%, which is 22.82% higher than the first coreflooding result. Interestingly, the continuous oil production is observed for LS-CO_2 WAG, but not for the secondary LSWF and tertiary CO_2 injection. It is concluded that LS-CO_2 WAG process introduces the synergy of LSWF and immiscible CO_2 injection on oil production. In addition, the WAG process achieves an optimized sweep efficiency to maximize the synergy of LSWF and immiscible CO_2 injection.

NUMERICAL SIMULATIONS

Dang, Nghiem, Chen, Nguyen, and Nguyen (2013) tried to combine the LSWF and CO_2 WAG processes numerically and constructed one-dimensional sandstone model. The numerical model of LS-CO_2 WAG process incorporates aqueous reactions, mineral dissolution of calcite, and cation exchange. The study assumed the mechanism of LSWF as the wettability modification changing relative permeability. The wettability modification is assumed to be attributed to the cation exchange of Ca^{2+}. In the previous experimental studies, it is clearly observed that the CO_2 solubility in brine influences the performance of CO_2 WAG process. Once the injected CO_2 dissolves into water, the dissolved CO_2 dissociates to produce hydrogen ions, i.e., lower pH. These reactions are implemented in the numerical model of LS-CO_2 WAG process. The secondary waterflood is applied until 200 days. The performance of tertiary LS-CO_2 WAG process is compared with the conventional waterflood, LSWF, CGI, and conventional WAG process. The LS-CO_2 WAG process produces more oil than other processes. The secondary application of LSWF ahead of tertiary LS-CO_2 WAG produces more oil. The higher oil recovery of LS-CO_2 WAG is a result of combination of wettability modification and miscibility effects. Especially, the injection of CO_2 decreasing pH possibly increases the Ca^{2+} concentration of in situ brine. It is attributed to the calcite mineral dissolution in lower pH. The higher concentration of Ca^{2+} increasing cation exchange of

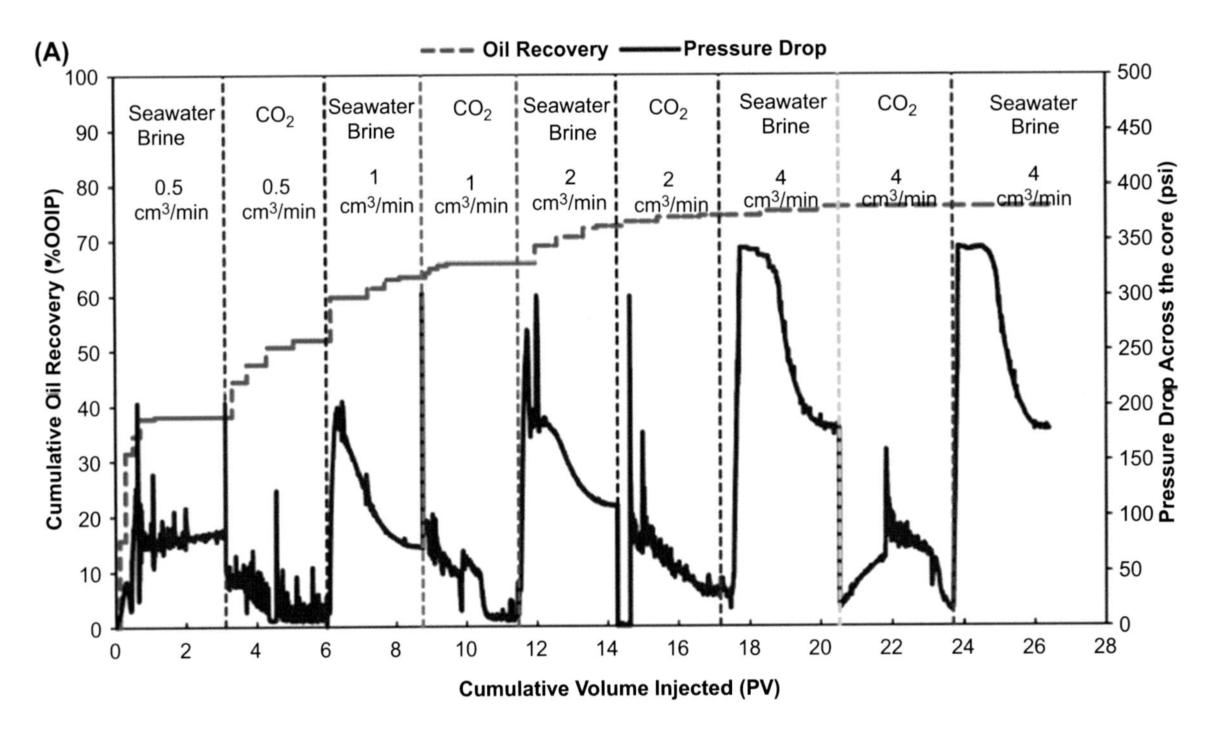

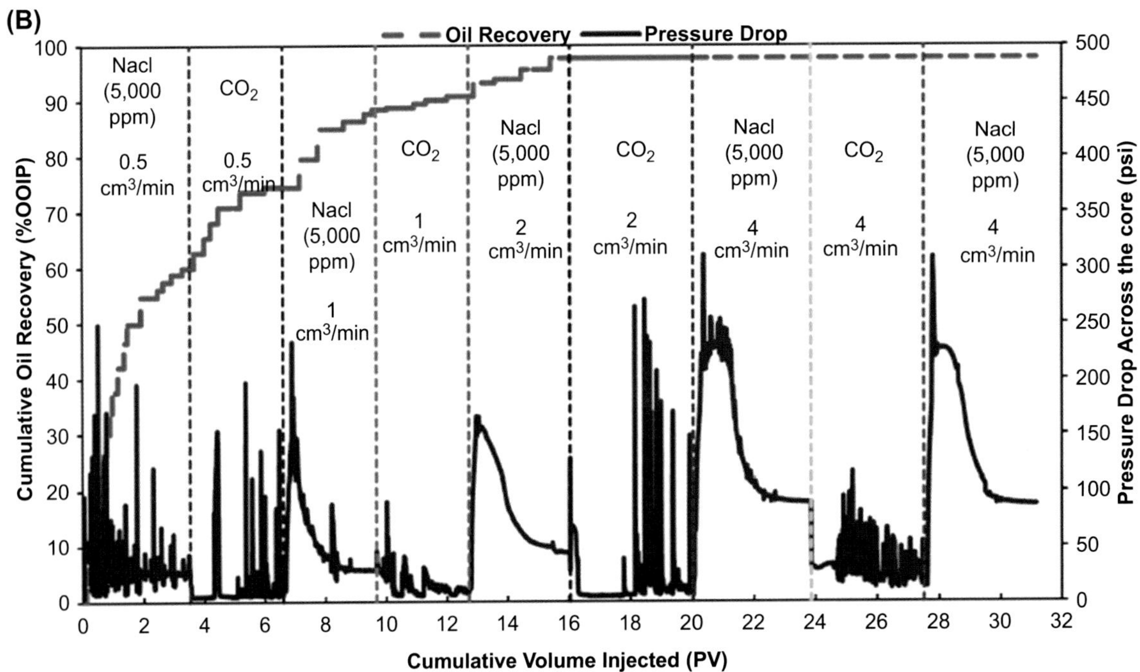

FIG. 5.5 The oil recovery and pressure drop across the aged core for the experiments of CO_2 WAG process using **(A)** seawater and **(B)** low-salinity water. (Credit: From Ramanathan, R., Shehata, A. M., & Nasr-El-Din, H. A. (2016). Effect of rock aging on oil recovery during water-alternating-CO_2 injection process: An interfacial tension, contact angle, Coreflood, and CT scan study. *Paper presented at the SPE improved oil recovery Conference, Tulsa, Oklahoma, USA, 11–13 April*. https://doi.org/10.2118/179674-MS.)

Ca^{2+} might modify the wettability additionally toward more water-wetness. Another study of Dang, Nghiem, Chen, Nguyen, and Nguyen (2014) reported more wettability modification effect with more cation exchange of Ca^{2+}. The comparison between LS-CO_2 WAG process with and without calcite reaction obviously confirms that higher recovery is obtained when the calcite mineral dissolves. Higher adherence of Ca^{2+} on rock by more cation exchange is observed because of calcite mineral dissolution. It is implied that CO_2 solubility in brine might reduce the amount of CO_2 to be miscible with oil. However, it can promote the wettability modification of LSWF mechanism. In addition, the LS-CO_2 WAG overcomes the delay in oil production, which is observed in CO_2 injection (Kulkarni & Rao, 2005). Comparing with CGI and conventional CO_2 WAG process, LS-CO_2 WAG shows the higher oil production rate as soon as CO_2 is injected (Fig. 5.6), which indicates the less delay in oil production. Dang et al. (2014) extended the numerical simulation of LS-CO_2 WAG process to the field-scaled assessment. Brugge benchmark field (Peters et al. 2010) is used for the deployment of LS-CO_2 WAG. Because the role of clay is important to simulate the wettability modification effect following LSWF mechanism in sandstone, the dispersed clay is additionally modeled. The reservoir model has the high uncertainty in clay distribution of reservoir. Considering the different facies and clay mapping, a number of geological realizations of clay distribution in reservoir are investigated. Considering the uncertainty, the LS-CO_2 WAG process provides the additional oil recovery from 4.5%–9% over conventional CO_2 WAG process.

The numerical studies (Al-Shalabi, Sepehrnoori, & Pope, 2014; Al-Shalabi, Sepehrnoori, & Pope, 2016) have reported the modeling of LS-CO_2 WAG process in carbonate rocks based on the numerical models of LSWF and CO_2 WAG process. The LS-CO_2 WAG process is assumed to be involved with the mechanism of LSWF and immiscible/miscible mechanisms of CO_2 WAG. It adapts the modeling of LSWF mechanism as empirical approach of wettability modification (Al-Shalabi, Sepehrnoori, Delshad, & Pope, 2015). Because LSWF model considers only two-phase flow of oil and water, the empirical approach of wettability modification

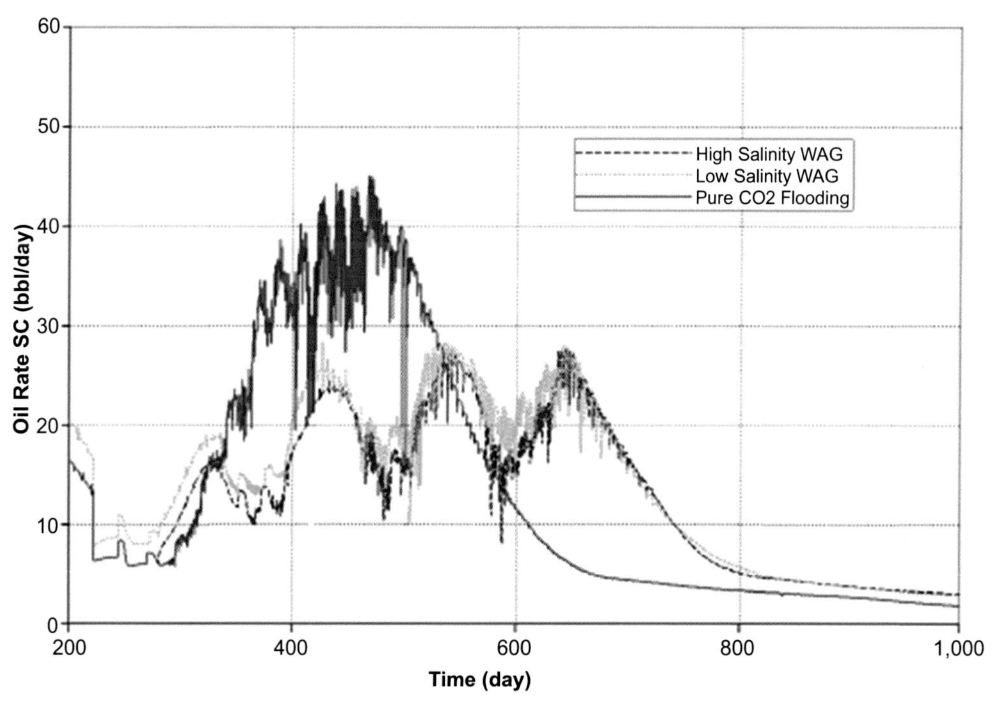

FIG. 5.6 Comparison of oil production rate of continuous CO_2 gas injection, conventional CO_2 water-alternating gas injection, and low salinity–assisted CO_2 water-alternating gas injection. (Credit: From Dang, C. T. Q., Nghiem, L. X., Chen, Z., Nguyen, N. T. B., & Nguyen, Q. P. (2014). CO_2 low salinity water alternating gas: A new promising approach for enhanced oil recovery. *Paper presented at the SPE improved oil recovery symposium, Tulsa, Oklahoma, USA, 12–16 April*. https://doi.org/10.2118/169071-MS.)

modeling changes the two-phase relative permeability as a function of salinity-dependent contact angle. Because the CO_2 WAG process incorporates the three-phase flow of oil, water, and gas, the Baker's model (Baker, 1988) is used for the three-phase relative permeability of mixed-wet formation. In the Baker's model, the three-phase oil relative permeability is obtained by the saturation-weighted interpolation between two-phase relative permeabilities of oil/water and oil/gas. The empirical approach of LSWF mechanism modifies the endpoint and Corey's exponent of oil corresponding only two-phase relative permeability of oil/water as well as residual oil saturation, then three-phase oil relative permeability of Baker's model is determined by incorporating the LSWF mechanism. In terms of modeling of CO_2 WAG process, the MMP of oil is calculated to determine whether the CO_2 WAG process is miscible or immiscible condition. The software of PVTsim, PVT simulator developed by Calsep Inc., and UTCOMP simulate the slim tube test to determine the MMP of oil at reservoir temperature of 248°F. The MMP is estimated as 3400 psi. The bubble point pressure is also calculated and determined as 3700 psi. It is unusual to have higher bubble point pressure than MMP of oil. Considering the uncertainty in the calculation of bubble point pressure and MMP, the MMP is determined to be equal to the bubble point pressure and has 3700 psi. The reservoir pressure of 4000 psi achieves the miscibility during the CO_2 WAG process. The possibility of asphaltene precipitation by depletion and CO_2 injection is investigated. The pressure-temperature diagram of phase behavior of asphaltene indicates no possibility of asphaltene precipitation even with CO_2 injection at reservoir temperature.

Before the simulation of LS-CO_2 EOR, the secondary seawater injection and tertiary CO_2 injection or conventional CO_2 WAG process using seawater are simulated. The seawater has the salinity of 43,610 ppm TDS. It has viscosity of 0.26 cp at reservoir temperature of 248°F. The oil and CO_2 gas have viscosities of 1.05 cp and 0.043688 cp at reservoir temperature. After the seawater injection, the equal amount of CO_2 is injected for tertiary CO_2 injection and conventional CO_2 WAG process. It is observed that either CO_2 injection or conventional CO_2 WAG recovers the additional oil over secondary injection. A number of designs of low salinity–augmented CO_2 injection are simulated with varying injection scheme and salinity. The low-salinity water is prepared by diluting the seawater. The injection schemes of CO_2 include the CGI, simultaneous water and CO_2 gas injection (SWAG), constant WAG, and tapered WAG. The SWAG, constant WAG, and tapered WAG use the low-salinity water for the water injection. Regardless of injection schemes of CO_2, additional oil recovery is obtained over the secondary oil recovery. As confirmed in the result of SWAG simulation (Fig. 5.7), the increased tertiary oil recovery is attributed to the combining effects of using low-salinity water and CO_2. It is believed that the main contribution to

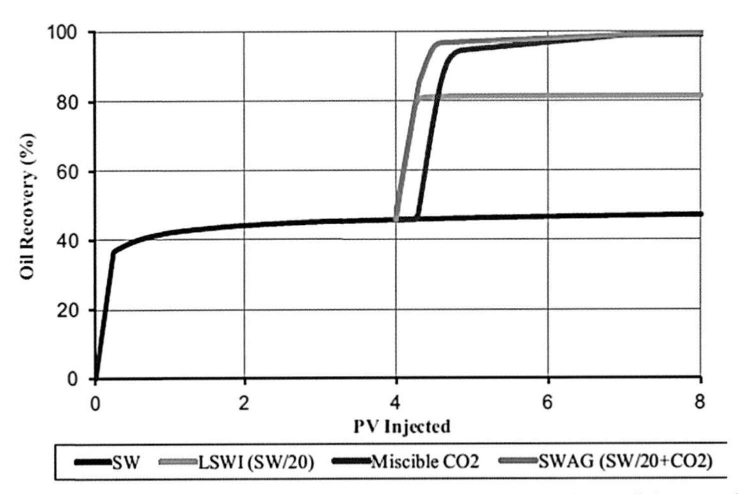

FIG. 5.7 Comparison of oil recovery between waterflood using seawater, low-salinity waterflood, miscible CO_2 injection, and simultaneous water and CO_2 injection using low-salinity water. (Credit: From Al-Shalabi, E. W., Sepehrnoori, K., & Pope, G. (2016). Numerical modeling of combined low salinity water and carbon dioxide in carbonate cores. *Journal of Petroleum Science and Engineering, 137*, 157–171. https://doi.org/10.1016/j.petrol.2015.11.021.)

the production of residual oil is the miscibility effect by CO_2 injection. The low-salinity water boosts the oil production rate by controlling oil relative permeability. Further simulations of SWAG using low-salinity water validate the synergetic effects of LS-CO_2 EOR. In addition, the fractional flow analysis of SWAG is performed. The analysis suggests that the SWAG using low-salinity water is a stable process and reduces the solvent addition to achieve an optimal condition.

Hamouda and Pranoto (2016) numerically investigated the LS-CO_2 WAG process in terms of oil production and geochemical analysis. The numerical model of LS-CO_2 WAG process refers the previous study of LSWF in chalk (Hamouda & Maevskiy, 2014) and applies the CO_2 injection into the LSWF process. Similar to the previous studies (Dang et al., 2013, 2014), the numerical model adapts a few geochemical reactions considering CO_2 solubility in water, aqueous reactions associated with the CO_2 dissolution in brine, calcite mineral reaction, and cation exchange. Because the wettability modification of LSWF mechanism is assumed to follow MIE theory, cation exchange determining the equivalent fraction of Ca^{2+} contributes the modification of relative permeability. A number of low-salinity waters are prepared by diluting seawater. The LS-CO_2 WAG process is designed to follow the LSWF. During the LS-CO_2 WAG process, an increased oil recovery is observed. The geochemical analysis in terms of the equivalent fraction of Ca^{2+} and Na^+ is performed. The process of LSWF increases the equivalent fraction of Ca^{2+} and decreases the equivalent fraction of Na^+. The more adherence of Ca^{2+} on the rock surface indicates the wettability modification improving relative permeability. The succeeding LS-CO_2 WAG additionally increases the equivalent fraction of Ca^{2+} and decreases the equivalent fraction of Na^+. Although the increment in the equivalent fraction of Ca^{2+} is small, it implies the potential to achieve additional wettability modification effect by CO_2 injection. As explained in the previous studies (Dang et al., 2013, 2014), calcite mineral dissolution by CO_2 injection could lead to the more cation exchange and additional wettability modification effect.

CARBONATED WATER INJECTION

The carbonated water injection (CWI) is water-based CO_2 flood. When carbonated water (CW) comes into contacting with oil, the CO_2 dissolved in CW transports to the oil, resulting in oil viscosity reduction and swelling (Sohrabi, Emadi, Farzaneh, & Ireland, 2015). These effects depend on the CO_2 solubility of brine. The potential of CWI has been investigated for decades. In the 1960s, it was firstly deployed in K&S project by ORCO (Oil Recovery Cooperation) in the Dewey-Bartlesville Field, Washington County, Northeast Oklahoma, for commercial production (Hickok, Christensen, & Ramsay, 1960). Hickok and Ramsay (1962) reported that CWI produced 43% of additional oil in K&S project and Project 33. A number of studies have tried to visualize the CWI phenomenon in glass micromodels or investigate the geochemical reactions associated with CWI process (Kechut, Sohrabi, & Jamiolahmady, 2011; Nunez, Vaz, Koroishi, Vidal Vargas, & Trevisan, 2017; Sohrabi et al., 2015). These studies confirmed oil swelling, oil viscosity reduction, the generation of CO_2-enriched gas, and mineral dissolution changing porosity and permeability during CWI. Moreover, the studies (Kechut et al., 2011; Sohrabi et al., 2015) have conducted coreflooding experiments and numerical simulations to observe whether oil production is enhanced or not.

Recently, the experimental attempts of combination between CWI and LSWF have been investigated (Kilybay et al., 2016, 2017). The low salinity–augmented CWI (LS-CWI) is proposed to recovery oil from oil-wet carbonate reservoir. Kilybay et al. (2016) carried out the coreflooding of LS-CWI process to estimate the EOR potential. The comprehensive experiments of IFT measurement, phase behavior test of microemulsion, NMR test, ζ-potential measurement, and ICP analysis investigate the understanding of rock-oil-brine interaction during LS-CWI in carbonate rocks. Before the implementation of LS-CWI into coreflooding, the optimized low-salinity water or smart water is evaluated. Five different brines are used such as deionized water, seawater, low-salinity water as 10-times-diluted seawater, modified seawater by adding 4-times sulfate concentration, and low-salinity water by adding 4-times sulfate concentration. The degassed and dewatered, and filtered oil samples are obtained from the oil reservoir of Abu Dhabi.

The microemulsion test using crude oil and brines is carried out at different pH conditions. The interface of oil and brine is stable, and no microemulsion is formed between oil and different brines at 95°C. The IFTs between the oil and brines of deionized water and seawater are measured by controlling the pH from 1 to 7. At pH around 1, the IFT is reduced down to 12 dyne/cm for all brines. However, the IFT is not enough to provide remarkable change to the oil production. The PALS technique measures the ζ-potentials of two different systems of brine-rock and brine-rock-oil. For the deionized brine, ζ-potentials of both systems show negative values. Basically, the carbonate rock surface is positively charged. In the test, the system of carbonated rock-deionized water shows the ζ-potential of −2.27 mV. Because the oil has negatively charged acidic molecules, the addition of crude oil into the brine-rock system changes the ζ-potential toward more negative. The system of carbonated rock-deionized water-crude oil has −37.9 mV. Following the double layer expansion and ion exchange phenomena, the LSWF or smart waterflood should change the surface charge of the system to positive value. The seawater, low-salinity water, modified seawater, and modified low-salinity water show the negative surface potentials of rock-brine-crude oil system. However, only modified seawater brine having four times higher concentration of sulfate shows the positive potential in the system of rock-brine-oil. Based on the ζ-potential measurement, the modified seawater is determined as the smart water and low-salinity water. Using the smart water, the carbonated water, which is fully CO_2-saturated water, is prepared at the experimental condition of coreflooding, 350 psi and 100°C. Neglecting the effect of salinity of water, the CO_2 solubility in water is referred from the work of Wiebe and Gaddy (1940). At the experimental condition, the 6 cc of CO_2 is determined to dissolve in 1 cc of water. Once the brine and estimated CO_2 are stored in the floating piston cylinder, the fluids are pressurized up to the 350 psi to dissolve CO_2 into brine. As the CO_2 dissolves in water, the pH of the brine decreases.

Three coreflooding experiments are designed to deploy in the sequence of waterflood using seawater, LSWF using modified seawater, and LS-CWI using modified seawater in carbonate cores. The permeability of cores ranges from 1.59 to 20.25 md. The tertiary recovery of LSWF varies from 9.4% to 0.66%. The LS-CWI increases the oil recoveries by 5.7%−13.9%. The recovery efficiency of LS-CWI is significantly higher compared with other processes. In addition to securing the wettability modification following the LSWF mechanism, it is suggested that three main mechanisms occur in the LS-CWI such as swelling and coalescence of trapped oil ganglia, local flow diversion, and oil viscosity reduction. Both the swelling and coalescence of trapped oil ganglia and oil viscosity reduction might improve the macroscopic sweep efficiency. In addition, the carbonated water has higher viscosity than normal water, and the higher viscosity of injecting carbonate water contributes to improve mobility ratio. When the carbonated water contacts oil, the mass transfer of CO_2 between carbonate water and oil reduces the oil viscosity. It is suggested that the mechanism of swelling and coalescence of trapped oil ganglia is dominant in light oil recovery because of higher solubility of CO_2 in lighter oil. For the heavy oil reservoir, the mechanism of viscosity reduction might be main contributor on the oil recovery controlling interfacial tension. Therefore, the LS-CWI is the promising EOR process in carbonate reservoir.

The NMR test and ICP analysis investigate the interaction of carbonate rock-brine. The NMR test estimates the porosity distribution of the three different cores, which are used in three coreflooding experiments. The porosity distributions of carbonate cores are estimated before and after coreflooding. After coreflooding test, the overall porosities are increased by 0.5%, 1.95%, and 2.31%, respectively. The increase in porosity is attributed to the carbonate rock dissolution. Mostly, the carbonate mineral dissolution increases the distribution of macroscopic pore. Some reduction in the distribution of microscopic pore is the result of precipitation of sulfate ions. As well as the injection of different ionic composition of brine, the mineral dissolution also changes the concentration of in situ and effluent brines. The ICP analysis measures the effluent concentrations of potential-determining ions of Mg^{2+}, $SO_4{}^{2-}$, and Ca^{2+}. The effluent concentration of $SO_4{}^{2-}$ is determined to be less than injecting concentration of $SO_4{}^{2-}$. The reduced concentration of $SO_4{}^{2-}$ implies the precipitation and deposition of sulfate-associated mineral or adsorption on the rock surface. The interpretation agrees with the observation of increasing distribution of microscopic pore from NMR test. In addition, higher concentration of Ca^{2+} is observed in the effluent brine than injecting brine because of carbonate mineral dissolution. The observation also corresponds to the increasing distribution of macroscopic pore of NMR test. In terms of concentration of Mg^{2+}, the consistent interpretation is hardly drawn. Kilybay et al. (2017) reported the further analysis of rock dissolution using the carbonate rock powders at various pH conditions. As the pH of brine decreases, the decreased weight of carbonate rock powders indicates mineral dissolution. These studies (Kilybay et al., 2016, 2017) have demonstrated the significant EOR potential of LS-CWI process in carbonate rocks and interactions in the rock-brine-oil contributing the EOR potential.

A couple of numerical studies (Lee, Jeong, & Lee, 2017; Lee & Lee, 2017) have investigated the LS-CWI and observed the synergy of mechanism of LSWF and CWI. Main contributions of LS-CWI on the EOR potential are the oil viscosity reduction, oil swelling, and wettability modification. In addition, the other studies (Lee, Kim, & Lee, 2017a, 2017b) have advanced the

LS-CWI process introducing the polymer flood. The hybrid polymer-assisted LS-CWI can enhance oil production maximizing sweep efficiency and improving displacement efficiency.

REFERENCES

Al-Shalabi, E. W., Sepehrnoori, K., Delshad, M., & Pope, G. (2015). A novel method to model low-salinity-water injection in carbonate oil reservoirs. *SPE Journal, 20*(05), 1154–1166. https://doi.org/10.2118/169674-PA.

Al-Shalabi, E. W., Sepehrnoori, K., & Pope, G. A. (2014). Modeling the combined effect of injecting low salinity water and carbon dioxide on oil recovery from carbonate cores. In *Paper presented at the international petroleum technology conference, Kuala Lumpur, Malaysia, 10–12 December*. https://doi.org/10.2523/IPTC-17862-MS.

Al-Shalabi, E. W., Sepehrnoori, K., & Pope, G. (2016). Numerical modeling of combined low salinity water and carbon dioxide in carbonate cores. *Journal of Petroleum Science and Engineering, 137*, 157–171. https://doi.org/10.1016/j.petrol.2015.11.021.

Aleidan, A. A., & Mamora, D. D. (2010). SWACO$_2$ and WACO$_2$ efficiency improvement in carbonate cores by lowering water salinity. In *Paper presented at the Canadian unconventional resources and international petroleum conference, Calgary, Alberta, Canada, 19–21 October*. https://doi.org/10.2118/137548-MS.

Bai, J.-M., Fan, W.-Y., Nan, G.-Z., Li, S.-P., & Bao-Shi, Y. (2010). Influence of interaction between heavy oil components and petroleum sulfonate on the oil–water interfacial tension. *Journal of Dispersion Science and Technology, 31*(4), 551–556. https://doi.org/10.1080/01932690903167475.

Baker, L. E. (1988). Three-phase relative permeability correlations. In *Paper presented at the SPE Enhanced oil recovery symposium, Tulsa, Oklahoma, 16–21 April*. https://doi.org/10.2118/17369-MS.

Caudle, B. H., & Dyes, A. B. (1958). *Improving miscible displacement by gas-water injection*. Society of Petroleum Engineers.

Chang, Y.-B., Coats, B. K., & Nolen, J. S. (1998). A compositional model for CO_2 floods including CO_2 solubility in water. *SPE Reservoir Evaluation and Engineering, 1*(02), 155–160. https://doi.org/10.2118/35164-PA.

Christensen, J. R., Stenby, E. H., & Skauge, A. (2001). Review of WAG field experience. *SPE Reservoir Evaluation and Engineering, 4*(02), 97–106. https://doi.org/10.2118/71203-PA.

Dang, C. T. Q., Nghiem, L. X., Chen, Z., Nguyen, Q. P., & Nguyen, N. T. B. (2013). State-of-the art low salinity waterflooding for enhanced oil recovery. In *Paper presented at the SPE Asia pacific oil and gas conference and exhibition, Jakarta, Indonesia, 22–24 October*. https://doi.org/10.2118/165903-MS.

Dang, C. T. Q., Nghiem, L. X., Chen, Z., Nguyen, N. T. B., & Nguyen, Q. P. (2014). CO_2 low salinity water alternating gas: A new promising approach for enhanced oil recovery. In *Paper presented at the SPE improved oil recovery symposium, Tulsa, Oklahoma, USA, 12–16 April*. https://doi.org/10.2118/169071-MS.

Hamouda, A. A., & Maevskiy, E. (2014). Oil recovery mechanism(s) by low salinity brines and their interaction with chalk. *Energy and Fuels, 28*(11), 6860–6868. https://doi.org/10.1021/ef501688u.

Hamouda, A. A., & Pranoto, A. (2016). Synergy between low salinity water flooding and CO_2 for EOR in chalk reservoirs. In *Paper presented at the SPE EOR conference at oil and gas West Asia, Muscat, Oman, 21–23 March*. https://doi.org/10.2118/179781-MS.

Harvey, A. H., & Prausnitz, J. M. (1989). Thermodynamics of high-pressure aqueous systems containing gases and salts. *AIChE Journal, 35*(4), 635–644. https://doi.org/10.1002/aic.690350413.

Hickok, C. W., Christensen, R. J., & Ramsay, H. J., Jr. (1960). Progress review of the K&S carbonated waterflood project. *Journal of Petroleum Technology, 12*(12), 20–24. https://doi.org/10.2118/1474-G.

Hickok, C. W., & Ramsay, H. J., Jr. (1962). Case histories of carbonated waterfloods in dewey-bartlesville field. In *Paper presented at the SPE secondary recovery symposium, Wichita Falls, Texas, 7–8 May*. https://doi.org/10.2118/333-MS.

Hjelmeland, O. S., & Larrondo, L. E. (1986). Experimental investigation of the effects of temperature, pressure, and crude oil composition on interfacial properties. *SPE Reservoir Engineering, 1*(04), 321–328. https://doi.org/10.2118/12124-PA.

Holm, W. L. (1987). Evolution of the carbon dioxide flooding processes. *Journal of Petroleum Technology, 39*(11), 1337–1342. https://doi.org/10.2118/17134-PA.

Holm, L. W., & O'Brien, L. J. (1971). Carbon dioxide test at the mead-strawn field. *Journal of Petroleum Technology, 23*(04), 431–442. https://doi.org/10.2118/3103-PA.

Jarrell, P. M., Fox, C. E., Stein, M. H., & Webb, S. L. (2002). *Practical aspects of CO_2 flooding, SPE monograph series*. Richardson, Tex: Henry L. Doherty Memorial Fund of AIME, Society of Petroleum Engineers.

Jiang, H., Nuryaningsih, L., & Adidharma, H. (2010). The effect of salinity of injection brine on water alternating gas performance in tertiary miscible carbon dioxide flooding: Experimental study. In *Paper presented at the SPE western regional meeting, Anaheim, California, USA, 27–29 May*. https://doi.org/10.2118/132369-MS.

Kechut, N. I., Sohrabi, M., & Jamiolahmady, M. (2011). Experimental and numerical evaluation of carbonated water injection (CWI) for improved oil recovery and CO_2 storage. In *Paper presented at the SPE EUROPEC/EAGE annual conference and exhibition, Vienna, Austria, 23–26 May*. https://doi.org/10.2118/143005-MS.

Kilybay, A., Ghosh, B., Chacko Thomas, N., & Aras, P. (2016). Hybrid EOR technology: Carbonated water and smart water improved recovery in oil wet carbonate formation. In *Paper presented at the SPE annual Caspian technical conference & exhibition, Astana, Kazakhstan, 1–3 November*. https://doi.org/10.2118/182567-MS.

Kilybay, A., Ghosh, B., Chacko Thomas, N., & Sulemana, N. T. (2017). Hybrid EOR technology: Carbonated water-smart water flood improved recovery in oil wet carbonate formation: Part-II. In *Paper presented at the SPE oil and gas India conference and exhibition, Mumbai, India, 4–6 April.* https://doi.org/10.2118/185321-MS.

Kulkarni, M. M., & Rao, D. N. (2005). Experimental investigation of miscible and immiscible water-alternating-gas (WAG) process performance. *Journal of Petroleum Science and Engineering, 48*(1), 120. https://doi.org/10.1016/j.petrol.2005.05.001.

Kumar, H. T., Shehata, A. M., & Nasr-El-Din, H. A. (2016). Effectiveness of low-salinity and CO_2 flooding hybrid approaches in low-permeability sandstone reservoirs. In *Paper presented at the SPE Trinidad and Tobago section energy resources conference, Port of Spain, Trinidad and Tobago, 13–15 June.* https://doi.org/10.2118/180875-MS.

Lake, L. W. (1989). *Enhanced oil recovery.* Englewood Cliffs, N.J.: Prentice Hall.

Langston, M. V., Hoadley, S. F., & Young, D. N. (1988). Definitive CO_2 flooding response in the SACROC unit. In *Paper presented at the SPE Enhanced oil recovery symposium, Tulsa, Oklahoma, 16–21 April.* https://doi.org/10.2118/17321-MS.

Lee, J. H., Jeong, M. S., & Lee, K. S. (2017). Geochemical modelling of carbonated low salinity water injection CLSWI to improve wettability modification and oil swelling in carbonate reservoir. In *Paper presented at the SPE Latin America and Caribbean mature fields symposium, Salvador, Bahia, Brazil, 15–16 March.* https://doi.org/10.2118/184915-MS.

Lee, J. H., Kim, G.-W., & Lee, K. S. (2017a). Low pH and salinity induced wettability modification of hybrid EOR as polymer-assisted carbonated low salinity waterflood in calcite cemented sandstone reservoir. In *Paper presented at the SPE/IATMI Asia pacific oil & gas conference and exhibition, Jakarta, Indonesia, 17–19 October.* https://doi.org/10.2118/186411-MS.

Lee, J. H., Kim, T. H., & Lee, K. S. (2017b). Hybrid CO_2 EOR using polymer-assisted carbonated low salinity waterflood to improve CO_2 deliverability and mobility. *Greenhouse Gases: Science and Technology, 8*(3), 444–461. https://doi.org/10.1002/ghg.1752.

Lee, J. H., & Lee, K. S. (2017). Enhanced wettability modification and CO_2 solubility effect by carbonated low salinity water injection in carbonate reservoirs. *Journal of Chemistry, 2017*, 10. https://doi.org/10.1155/2017/8142032.

Martin, J. W. (1951). Additional oil production through flooding with carbonated water. *Producers Monthly, 15*(9), 18–22.

Nunez, R., Vaz, R. G., Koroishi, E. T., Vidal Vargas, J. A., & Trevisan, O. V. (2017). Investigation of dissolution effects on dolomite porous media under carbonated water injection CWI. In *Paper presented at the Abu Dhabi international petroleum exhibition & conference, Abu Dhabi, UAE, 13–16 November.* https://doi.org/10.2118/188601-MS.

Orr, F. M., & Taber, J. J. (1984). Use of carbon dioxide in enhanced oil recovery. *Science, 224*(4649), 563.

Peters, L., Arts, R., Brouwer, G., Geel, C., Cullick, S., Lorentzen, R. J., et al. (2010). Results of the Brugge benchmark study for flooding optimization and history matching. *SPE Reservoir Evaluation and Engineering, 13*(03), 391–405. https://doi.org/10.2118/119094-PA.

Ramanathan, R., Shehata, A. M., & Nasr-El-Din, H. A. (2015). Water alternating CO_2 injection process – does modifying the salinity of injected brine improve oil recovery? In *Paper presented at the OTC Brasil, Rio de Janeiro, Brazil, 27–29 October.* https://doi.org/10.4043/26253-MS.

Ramanathan, R., Shehata, A. M., & Nasr-El-Din, H. A. (2016). Effect of rock aging on oil recovery during water-alternating-CO_2 injection process: An interfacial tension, contact angle, coreflood, and CT scan study. In *Paper presented at the SPE improved oil recovery conference, Tulsa, Oklahoma, USA, 11–13 April.* https://doi.org/10.2118/179674-MS.

Rogers, J. D., & Grigg, R. B. (2001). A literature analysis of the WAG injectivity abnormalities in the CO_2 process. *SPE Reservoir Evaluation and Engineering, 4*(05), 375–386. https://doi.org/10.2118/73830-PA.

Setschenow, J. (1889). Über die Konstitution der Salzlösungen auf Grund ihres Verhaltens zu Kohlensäure. *In Zeitschrift für Physikalische Chemie, 4*, 117–125.

Slobod, R. L., & Koch, H. A., Jr. (1953). High-pressure gas injection - mechanism of recovery increase. In *Paper presented at the drilling and production practice, New York, New York, 1 January.*

Sohrabi, M., Emadi, A., Farzaneh, S. A., & Ireland, S. (2015). A thorough investigation of mechanisms of enhanced oil recovery by carbonated water injection. In *Paper presented at the SPE annual technical conference and exhibition, Houston, Texas, USA, 28–30 September.* https://doi.org/10.2118/175159-MS.

Søreide, I., & Whitson, C. H. (1992). Peng-Robinson predictions for hydrocarbons, CO_2, N_2, and H_2S with pure water and NaCI brine. *Fluid Phase Equilibria, 77*, 217–240. https://doi.org/10.1016/0378-3812(92)85105-H.

Stalkup, F. I. (1983). *Miscible displacement, monograph/society of petroleum engineers.* New York: Henry L. Doherty Memorial Fund of AIME, Society of Petroleum Engineers of AIME.

Teklu, T. W., Alameri, W., Graves, R. M., Kazemi, H., & Al-sumaiti, A. M. (2014). Low-salinity water-alternating-CO_2 flooding enhanced oil recovery: Theory and experiments. In *Paper presented at the Abu Dhabi international petroleum exhibition and conference, Abu Dhabi, UAE, 10–13 November.* https://doi.org/10.2118/171767-MS.

Teklu, T. W., Alameri, W., Graves, R. M., Kazemi, H., & AlSumaiti, A. M. (2016). Low-salinity water-alternating-CO_2 EOR. *Journal of Petroleum Science and Engineering, 142*, 101–118. https://doi.org/10.1016/j.petrol.2016.01.031.

Whorton, L. P., Brownscombe, E. R., & Dyes, A. B. (1952). *Method for producing oil by means of carbon dioxide.* Atlantic Refining Co.

Wiebe, R., & Gaddy, V. L. (1940). The solubility of carbon dioxide in water at various temperatures from 12 to 40° and at pressures to 500 atmospheres. Critical phenomena. *Journal of the American Chemical Society, 62*(4), 815–817. https://doi.org/10.1021/ja01861a033.

Zick, A. A. (1986). A combined condensing/vaporizing mechanism in the displacement of oil by enriched gases. In *Paper presented at the SPE annual technical conference and exhibition, New Orleans, Louisiana, 5–8 October.* https://doi.org/10.2118/15493-MS.

CHAPTER 6

Hybrid Thermal Recovery Using Low-Salinity and Smart Waterflood

ABSTRACT

This chapter discusses the applications of hybrid process by coupling low-salinity waterflood and thermal recovery methods, i.e., low salinity-augmented thermal recovery. The thermal recovery methods include hot water injection and steam injection. Research studies of experimental and numerical simulation have evaluated the performances of hot water injection and steam injection on heavy oil production by controlling water chemistry, especially salinity. The synergy of low salinity–augmented thermal recovery on heavy oil production is discussed in this chapter.

The thermal recovery method supplies the thermal energy to reservoirs through two processes: (1) process in which the heat is produced at the surface and (2) process in which the heat is created in the formation (Burger, Sourieau, & Combarnous, 1985; Latil, 1980). The first process injects the heated fluids into the target reservoirs, while the second process injects the reactants occurring during exothermic reactions in the reservoir formations. Because the first process loses the heat to the surrounding formations as the injecting fluid flows, the performance is dependent on its thermal efficiency. Because the second process exactly releases the heat at the target zone where the viscous oil is to be mobilized, it has negligible risk of the heat loss. The first process of thermal recovery method includes hot fluid injection, hot water injection, steam injection, cyclic steam injection, and steam-assisted gravity drainage. The second process corresponds to the in-situ combustion. There are several variants of in-situ combustion such as forward dry in-situ combustion and wet combustion or partially quenched combustion. In addition, there are advanced technologies of thermal EOR including the toe-to-heel air injection. Of that, the first process has been proposed to be candidate for the thermal EOR process for the hybrid low salinity–augmented thermal recovery method. Prior to the description of low salinity–augmented thermal recovery method, the benefits of thermal recovery methods on the oil production are briefly summarized. Overall explanation of the thermal recovery method is discussed in the references (Burger et al., 1985; Latil, 1980).

Thermal recovery methods modify the rock and fluid properties and therefore, the dynamic behavior of the fluid in a porous media. The heat from the thermal recovery methods potentially affects hydrodynamic properties (liquid viscosity and relative permeability), thermal and thermal dynamic properties (thermal expansion, thermal capacity, thermal conductivity, and latent heat of vaporization), and chemical reactions. The effects of temperature on the hydrodynamic properties and thermal dynamic properties are discussed. In addition, the heat loss, which is an important factor in thermal efficiency of the first process, is also discussed.

HYDRODYNAMIC PROPERTIES

The liquid viscosity is highly affected by the heat. The viscosity decreases with an increase in temperature. Most liquids suffer the exponential relationship (Seeton, 2006) between temperature and viscosity rather than linear form (Fig. 6.1). The more viscous the fluid, the more sensitive it is to the temperature change. Because higher temperature makes both slightly less viscous water and much lesser viscous oil, the improved viscosity contrast favors relatively the oil flow rather than water flow. This contribution by heat enables to deploy the exploitation of thermal recovery method in mainly heavy oil reservoirs. In addition, the dissolution of gases such as CO_2 can reduce the viscosity of liquid hydrocarbons. During the in-situ combustion, a large quantity of CO_2 forms in reservoirs. The CO_2 can dissolve in oil at high pressure condition and the mobility of oil improves. For the gas viscosity, kinetic theory of ideal gases explains that the dynamic viscosity of ideal gases is proportional to the square root of the absolute temperature and not sensitive to the pressure. However, there is some discrepancy between real gases and ideal gases behaviors. The real gases viscosity tends to increase with an increase in pressure. The sensitivity of viscosity to temperature is higher in real gases than ideal gas behaviors.

Hybrid Enhanced Oil Recovery using Smart Waterflooding. https://doi.org/10.1016/B978-0-12-816776-2.00006-4

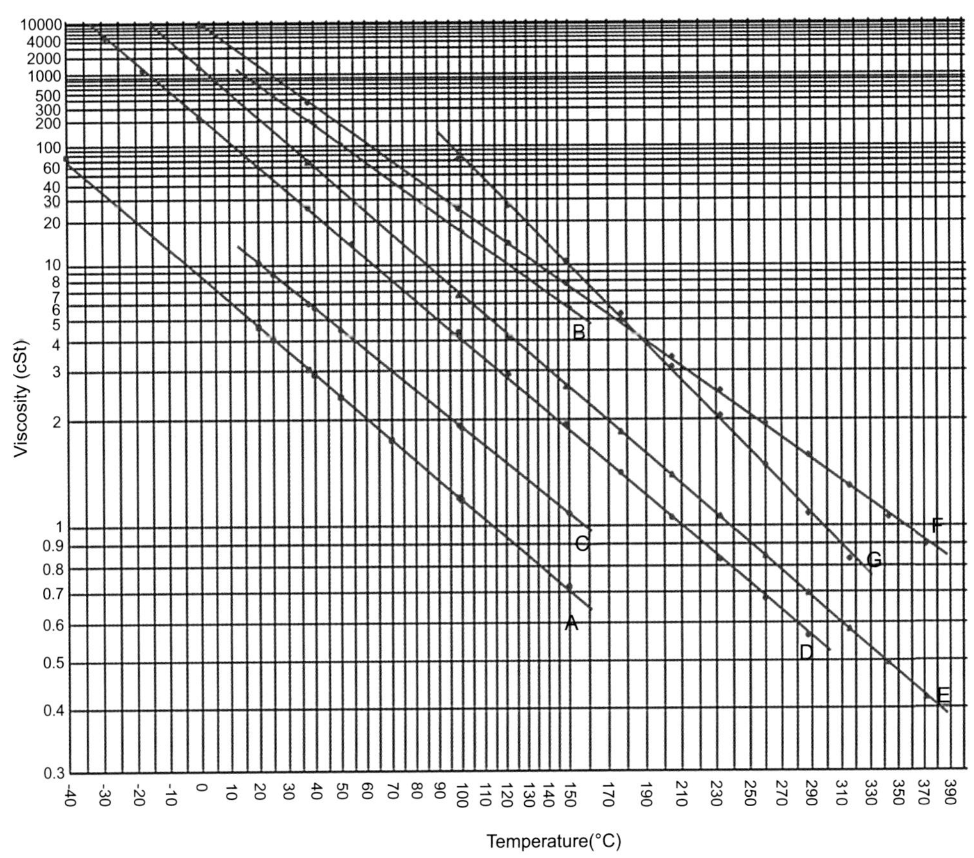

FIG. 6.1 The exponential relationship between viscosity and temperature for various oil samples. (Credit: from Seeton, C. J. (2006). Viscosity–temperature correlation for liquids. *Tribology Letters*, *22*(1), 67–78. https://doi.org/10.1007/s11249-006-9071-2.)

The dependency of relative permeability to temperature has been exploited by various experiments (Lo & Mungan, 1973; Poston, Ysrael, Hossain, & Montgomery, 1970; Qin, Wu, Liu, Zhao, & Yuan, 2018; Schembre, Tang, & Robert Kovscek, 2005; Weinbrandt, Ramey, & Casse, 1975). As the irreducible water saturation increases, residual oil saturation decreases, and the endpoint of water relative permeability increases as the temperature increases (Fig. 6.2). However, the chemical reactions or vaporization and condensing phenomena of gaseous components during in-situ combustion could lead to unexpected results.

THERMAL AND THERMAL DYNAMIC PROPERTIES

The transport of thermal energy in reservoirs is affected by a number of factors including thermal expansion, thermal capacity, and thermal conductivity. The water, oil, and rock exhibit the thermal expansion, and their coefficients of thermal expansion are on the orders of 10^{-3}, 10^{-4}, and $10^{-5}°C^{-1}$. The thermal expansion of rock affects the expulsion of oil from pore space. In terms of thermal capacity per unit mass, i.e., specific heat, the specific heats of the liquids, gases, and rocks increase with an increase in the temperature. In addition, the specific heats vary according to the composition. The thermal conductivity of gases increases with temperature, while that of most liquids and solids decreases slightly with temperature. Because the reservoir system is the combination of rock, liquids, and gases, the equivalent thermal conduction of reservoir system is defined considering the porosity, saturation of fluid, and thermal conductivity. The last factor, the latent heat of vaporization, reduces the thermal energy to be used for mechanisms of thermal recovery methods.

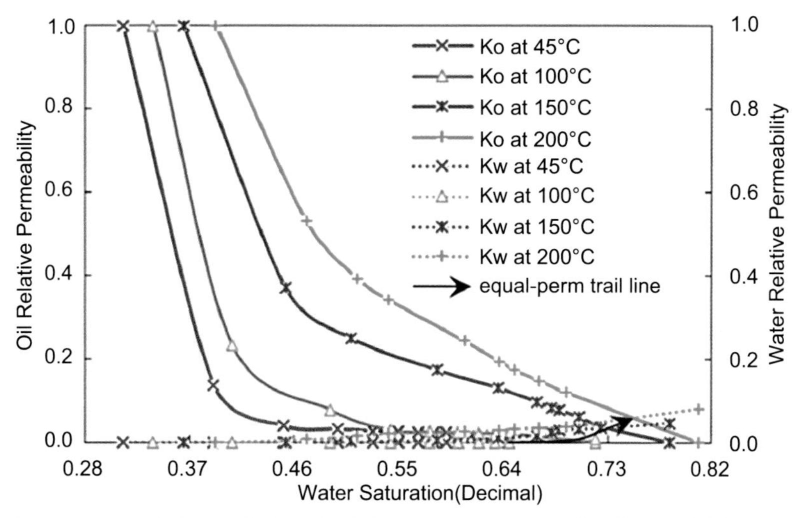

FIG. 6.2 Relative permeability modification due to temperature change. (Credit: Qin, Y., Wu, Y., Liu, P., Zhao, F., & Yuan, Z. (2018). Experimental studies on effects of temperature on oil and water relative permeability in heavy-oil reservoirs. *Scientific Reports*, *8*(1), 12530. https://doi.org/10.1038/s41598-018-31044-x.)

HEAT LOSS

The heat loss to the surrounding formations is the major concern in the first process of thermal recovery method. The heat loss reduces the temperature of the hot zone except in the steam condensation zone and affects the thermal efficiency of the process. Therefore, the expectation of heat loss is of importance. Because of the heat loss, the conditions of thin reservoirs or the large well spacing between injector and producer are not appropriate for the hot fluid injection. While the hot fluid flows from the surface condition to the target zones through the injector, significant heat loss occurs. The injection well is surrounded by the colder formations or conditions. In addition, the part of enthalpy of fluid is lost by either temperature decreases of hot fluid or steam quality reduction. In these conditions, heat transfer occurs by conduction, convection, and radiation.

HOT WATER INJECTION

Low Salinity–Augmented Hot Water Injection

In hot fluid injection, preheated fluids are injected into relatively cold reservoirs. The injecting fluids are heated at the surface or well bore using well bore heaters such as down-hole heaters. The fluids can be water, air, natural gas, CO_2, solvent, etc. The hot fluid injection using hot water, i.e., hot water injection, is involved with the two-phase flow of oleic and aqueous phases. The front of hot water loses the heat to contacting reservoir zone and the transferred heat improves the fluid mobility in the heated zone. This process modifies the displacement efficiency of heated zone and increases the ultimate heavy oil recovery. In addition, the thermal expansion of oil contributes to the recovery by oil displacement.

Abass and Fahmi (2013) proposed the low salinity–augmented hot water injection or LS-hot water injection, and experimentally evaluated its heavy oil recovery from sandstone. The coreflooding experiments compare the performance of LS-hot water injection to the conventional waterflood, hot water injection, steam injection, and LSWF using a BAW field sandstone core and unconsolidated sand packs. Heavy oil from the BAW field and two additional heavy oils are used. The oil from the BAW field has the viscosity of 1700 cp at 65°C and other oils have viscosities of 700 cp at 35°C and higher than 1000 cp at 60°C. The high salinity brine of 24,000 ppm TDS and low salinity brine of 200 ppm TDS are prepared. Three displacement experiments with unconsolidated sand packs and one coreflooding experiment with a BAW field core are performed.

In the first displacement, a sand pack is saturated with the BAW field oil and is set to have temperature condition of 65°C. The low salinity water with 200 ppm TDS is used for the connate water. Conventional hot water injection and LS-hot water injection with 95°C are deployed in the sand pack system. When the water-cut of displacement reaches 0.9, secondary conventional hot water injection is converted to tertiary LS-hot water injection. Not only the reduction in water-cut, but also an increase in oil recovery is observed by switching from hot water injection to LS-hot water injection

(Fig. 6.3). The production of conventional hot water injection shows the contributions of thermal energy, i.e., heavy oil viscosity reduction and thermal expansion of oil, on the heavy oil production. During the LS-hot water injection, the additional increases of oil recovery by about 0.2 are attributed to the wettability modification following LSWF mechanism. The second displacement experiment also simulates the equivalent injection design of secondary hot water injection and tertiary LS-hot water injection. The hot water injection and LS-hot water injection are operated at 90°C. The sand pack with about 2944 md is saturated with another heavy oil. In contrast to the first displacement experiment, the connate water has high salinity of 15,000 ppm TDS. The salinity of conventional hot water injection also increases to a salinity of 24,000 ppm TDS. The secondary conventional hot water injection is ceased when the breakthrough of injecting high salinity water is observed. Then, the tertiary LS-hot water injection applies. The LS-hot water injection recovers the additional oil recovery up to 0.29 above the hot water injection. The two displacement experiments clearly demonstrate that the hot water injection process can introduce the LSWF effect by decreasing salinity of injection water. The third displacement experiment investigates the contribution of steam injection on the additional heavy oil production after secondary hot water injection and tertiary LS-hot water injection at 100°C. The connate water has a low salinity of 200 ppm TDS. The thermal expansion of oil and oil viscosity reduction by hot water injection result in the secondary oil recovery of about 0.22. The wettability modification effect during tertiary recovery of LS-hot water injection boosts the thermal contribution on heavy oil recovery up to 0.52. The steam injection at 250°C after LS-hot water injection still increases the oil recovery by about 0.08. The last coreflooding is

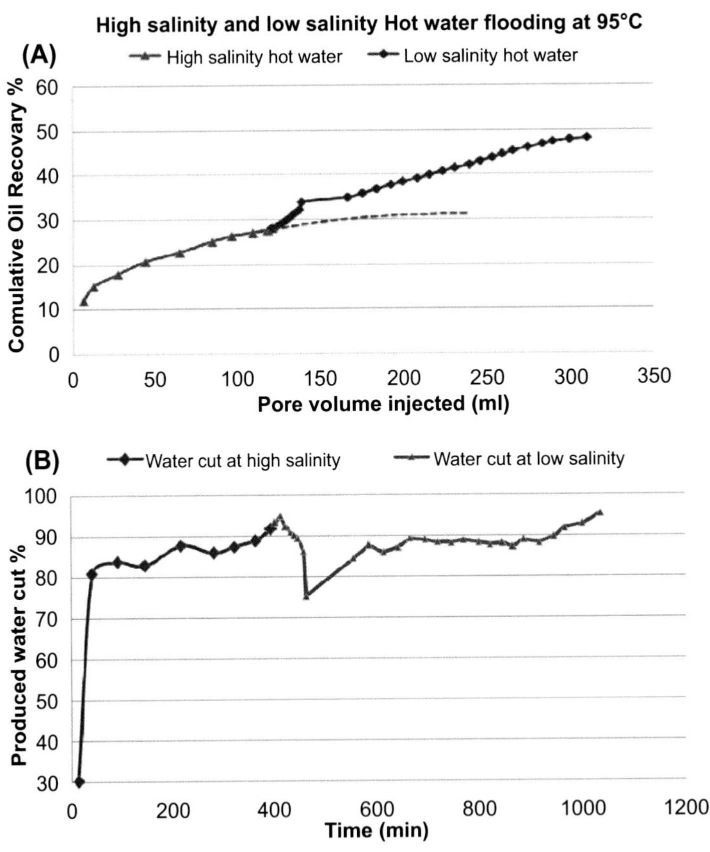

FIG. 6.3 Productions of secondary hot water injection and tertiary low salinity–augmented hot water injection: **(A)** oil production and **(B)** water-cut. (Credit: from Abass, E., & Fahmi, A. (2013). Experimental investigation of low salinity hot water injection to enhance the recovery of heavy oil reservoirs. *Paper presented at the North Africa technical conference and exhibition, Cairo, Egypt, 15–17 April*. https://doi.org/10.2118/164768-MS.)

simulated to confirm the effectiveness of LS-hot water injection after hot water injection. The observation of coreflooding corresponds to that of displacement experiments using sand packs. The reduction in watercut and an increase in heavy oil production are also observed during tertiary LS-hot water injection.

Based on these experimental observations, this study summarized the benefits of LS-hot water injection to enhance the heavy oil production from sandstone reservoirs. The hot water injection can be comparable to the steam injection by reducing the salinity of injecting brine in terms of intermediate heavy oil EOR potential in the BAW Field. The hybrid technology can introduce the thermal expansion, oil viscosity reduction, and wettability modification of LSWF effect. Although the replacement of steam injection by LS-hot water injection could save the cost of steam generation and transportation, it requires the additional cost of the water supply and desalination treatment.

Al-Saedi, Flori, and Brady (2018) investigated the effect of hot water on the performance of LSWF process and, experimentally, quantified the heavy oil production of LS-hot water injection. Target reservoir is the Kansas heavy oil reservoir in Midwestern reservoirs. The heavy oil in the field is hardly produced by natural depletion. It has the low temperature condition with high viscosity of oil. The oil sample has the viscosity of 600 cp at 20°C. The Berea sandstone core, which has permeability of about 100 md and porosity of 0.2, is used. The four sets of coreflooding test are carried out with different temperature conditions (70 and 90°C) and different connate waters. Formation water has the salinity of 97,500 ppm TDS, and modified formation water is prepared by increasing the concentration of Ca^{2+} by a factor of two. The formation water and modified formation water are used for the connate water. The low salinity water is manufactured by diluting the formation water by a factor of 100. The injection design is set to apply the secondary waterflood injecting formation water and tertiary LSWF. The slug of secondary or tertiary injections is injected as much as 2 PV.

The first coreflooding is operated at 70°C. The connate water is the formation water. The conventional waterflood produces the oil recovery of about 0.42 and tertiary LSWF produces the additional oil recovery of about 0.08. The experimental condition of second coreflooding is the same with the first coreflooding except for the temperature condition. The second coreflooding is simulated at 90°C to investigate the process of hot water injection. The secondary hot water injection injecting formation brine results in the oil recovery of about 0.45. The succeeding LS-hot water injection produces more oil recovery by about 0.1. In the comparison between two coreflooding tests, it is clearly observed that the LS-hot water injection introduces the thermal expansion and heavy oil viscosity reduction, and secures the wettability modification effect of LSWF mechanism. To assess the role of Ca^{2+} on the performance of LS-hot water injection, the third and fourth coreflooding tests are performed by changing connate water from formation water to modified formation water. The third coreflooding at 70°C and fourth coreflooding at 90°C show the higher effluent pH compared to the first and second tests during the injection of low salinity water. In LSWF condition, the high content of Ca^{2+} increases the pH regardless of temperature condition. However, LSWF leads to less improvement in oil recovery for both temperature conditions when connate water has higher Ca^{2+} concentration. It is determined that the increase in pH is not a necessary sign to improve the effect of LSWF in hot and low temperature conditions. In the study, a couple of conclusions are drawn for LS-hot water injection. Both the reducing salinity and increasing temperature contribute to the heavy oil production in sandstone reservoir. Controlling the chemistry of water could solve and supplement the limitations in the hot water injection process.

Studies (Lee, Jeong, & Lee, 2016; Mohammadi, 2017) have reported the numerical simulations of LS-hot water injection for heavy oil recovery. The numerical simulation employs the comprehensive geochemical reactions and temperature-dependent reduction of oil viscosity in the modeling of LS-hot water injection. The LS-hot water injection is compared to the LSWF in terms of geochemical reactions and oil viscosity. The wettability mechanism of LSWF process is assumed to be attributed to the cation exchange of Ca^{2+} and the assumption is applied to the simulation of LS-hot water injection. The wettability modification is modeled by the modification of relative permeability. The high temperature condition accelerates the dissolution of carbonate minerals. More dissolution of carbonate minerals generates the in-situ concentration of Ca^{2+}, potentially modifying wettability. Therefore, the temperature-dependent geochemical reactions of mineral dissolution and cation exchange enhance the wettability modification effect during LS-hot water injection compared to the LSWF. The significant oil viscosity reduction also attributes to the heavy oil production. Up to date, the numerical simulations of hot-LSWI are limited.

The process of hot water injection is competitive in relatively less viscous oil on the order of a few hundred centipoises in specific conditions. However, the high risk of low thermal efficiency is the barrier of hot water injection applications. Because of low thermal efficiency, extensive thermal energy is useless, and only a small fraction of energy is used to heat the target fluids.

STEAM INJECTION

Steam injection conventionally has several forms. The main steam processes are cyclic steam stimulation and steam flooding. Conventionally, the cyclic steam stimulation uses a single well for both injection and production, and steam flooding is designed with multiwells for injector and producer. The cyclic steam stimulation heats the oil and improves the mobilization of oil in the vicinity of the well. The heated oil is produced by a number of driving forces, including the reservoir pressure depletion, gravity, formation compaction, etc. The steam flooding injects the steam into the reservoir and creates steam zones, in which the oil mobility is affected. Because the steam has the tendency to migrate toward the upper portion of formation due to gravity segregation, the steam zones grow both laterally and vertically. The heated and mobilized oil in the steam zones is recovered through the producer. There are other variants of steam injection by modifying the two processes with additives addition, horizontal well, conductive heating using fractures, etc. to overcome the inherent risks or improves the efficiency.

The steam flooding shows the inherent low sweep efficiency due to early steam breakthrough and gravity override. It can be improved by adapting the water-alternating-steam process (WASP). The studies (Al-Saedi, Flori, Alkhamis, & Brady, 2018a,b) have proposed the low salinity–alternating steam flooding (LSASF) process to combine the WASP and LSWF. The experimental application of LSASF is evaluated to recover Kansas heavy oil from Berea sandstone cores. The four sets of coreflooding are conducted to investigate the tertiary recovery of LSASF. The steam temperature is 150°C, and the experimental condition of temperature is the room temperature. The high saline formation water has a salinity of 108,460 ppm TDS. The low salinity water is prepared by diluting the formation water by a factor of 100.

The first coreflooding is designed with the secondary conventional waterflood, one cycle of LSASF, and chasing LSWF. The secondary conventional water injecting formation water produces an oil recovery of about 42%. Then, the tertiary injection of LSASF follows the secondary injection. During the tertiary LSASF, the low salinity water injection and steam injection recover the additional oil recovery of 2.8% and 1.86%, respectively. The last chasing low salinity water injection after steam injection produces more oil recovery of 3.7%. The second and third coreflooding tests increase the cycle of LSASF process by two. The difference between the second and third coreflooding tests is the slug size injected. Both second and third coreflooding tests recover significant oil within the second cycle of LSASF process (Fig. 6.4A). The ultimate oil recoveries are 57.8% and 62.7% for the second and third coreflooding

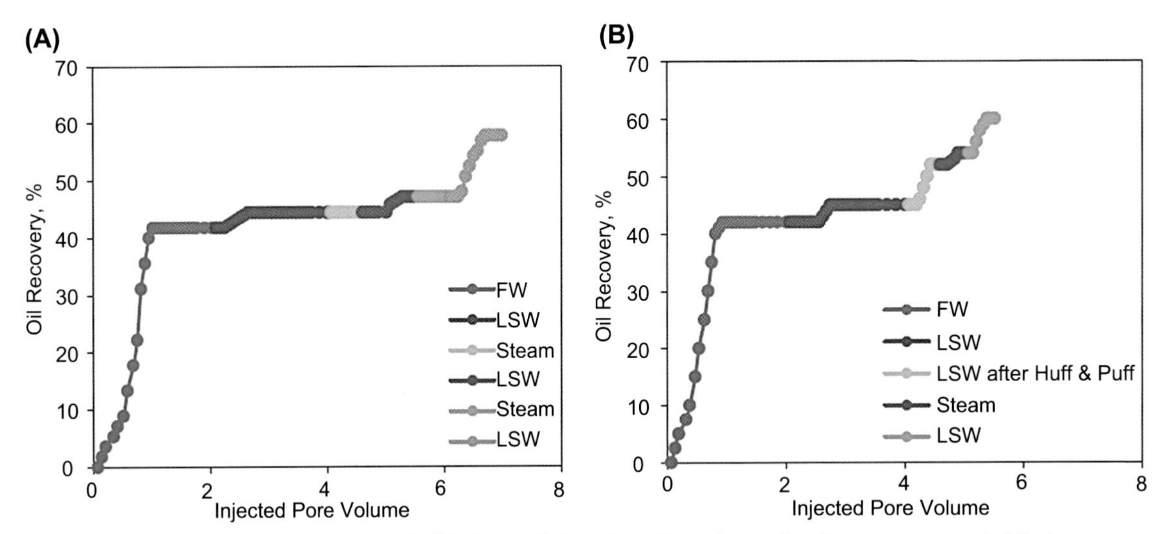

FIG. 6.4 Heavy oil recovery of **(A)** low salinity–alternating steam flooding process and **(B)** low salinity–alternating steam flooding with huff and puff process. (Credit: From Al-Saedi, H. N., Flori, R. E., Alkhamis, M., & Brady, P. V. (2018b). Coupling of low-salinity water flooding and steam flooding for sandstone unconventional oil reservoirs. *Natural Resources Research*, https://doi.org/10.1007/s11053-018-9407-2.)

experiments, respectively. In the last coreflooding, the huff and puff process using steam is implemented in tertiary recovery. The secondary recovery of conventional waterflood produces oil recovery of 42%, and the tertiary application produces the ultimate recovery of 60% (Fig. 6.4B). This experimental study clearly demonstrated the high potential of hybrid process of steam injection and LSWF to produce the heavy oil. The steam injection is efficient to reduce the heavy oil viscosity, and LSWF modifies the wettability toward water-wetness. The LSASF could secure both the advantages from steam injection and LSWF. In addition to the synergy, the LSASF could delay the production of steam and consume the less amount of steam to be injected. Therefore, the LSASF could increase both heavy oil EOR potential and economics.

REFERENCES

Abass, E., & Fahmi, A. (2013). Experimental investigation of low salinity hot water injection to enhance the recovery of heavy oil reservoirs. In *Paper presented at the North Africa technical conference and exhibition, Cairo, Egypt, 15–17 April.* https://doi.org/10.2118/164768-MS.

Al-Saedi, H. N., Flori, R. E., Alkhamis, M., & Brady, P. V. (2018a). Coupling low salinity water flooding and steam flooding for sandstone reservoirs; low salinity-alternating-steam flooding (LSASF). In *Paper presented at the SPE kingdom of Saudi Arabia annual technical symposium and exhibition, Dammam, Saudi Arabia, 23–26 April.* https://doi.org/10.2118/192168-MS.

Al-Saedi, H. N., Flori, R. E., Alkhamis, M., & Brady, P. V. (2018b). Coupling of low-salinity water flooding and steam flooding for sandstone unconventional oil reservoirs. *Natural Resources Research.* https://doi.org/10.1007/s11053-018-9407-2.

Al-Saedi, H. N., Flori, R. E., & Brady, P. V. (2018). Enhanced heavy oil recovery by thermal-different aqueous ionic solutions-low salinity water flooding. In *Paper presented at the SPE kingdom of Saudi Arabia annual technical symposium and exhibition, Dammam, Saudi Arabia, 23–26 April.* https://doi.org/10.2118/192179-MS.

Burger, J., Sourieau, P., & Combarnous, M. (1985). *Thermal methods of oil recovery, Institut français du pétrole publications.* Houston Paris: Gulf Pub. Co. Éditions Technip.

Latil, M. (1980). *Enhanced oil recovery, Institut français du pétrole publications.* Paris: Éditions Technip.

Lee, J. H., Jeong, M. S., & Lee, K. S. (2016). Thermo-mechanistic EOR process modelling in deploying low salinity hot water injection under carbonate reservoirs. In *Paper presented at the international petroleum technology conference, Bangkok, Thailand, 14–16 November.* https://doi.org/10.2523/IPTC-18660-MS.

Lo, H. Y., & Mungan, N. (1973). Effect of temperature on water-oil relative permeabilities in oil-wet and water-wet systems. In *Paper presented at the fall meeting of the society of petroleum engineers of AIME, Las Vegas, Nevada, 30 September–3 October.* https://doi.org/10.2118/4505-MS.

Mohammadi, A. H. (2017). *Heavy oil: Characteristics, production and emerging technologies, petroleum science and technology.* New York: Nova Publishers.

Poston, S. W., Ysrael, S., Hossain, A. K. M. S., & Montgomery, E. F., III. (1970). The effect of temperature on irreducible water saturation and relative permeability of unconsolidated sands. *Society of Petroleum Engineers Journal, 10*(02), 171–180. https://doi.org/10.2118/1897-PA.

Qin, Y., Wu, Y., Liu, P., Zhao, F., & Yuan, Z. (2018). Experimental studies on effects of temperature on oil and water relative permeability in heavy-oil reservoirs. *Scientific Reports, 8*(1), 12530. https://doi.org/10.1038/s41598-018-31044-x.

Schembre, J. M., Tang, G.-qing, & Robert Kovscek, A. (2005). Effect of temperature on relative permeability for heavy-oil diatomite reservoirs. In *Paper presented at the SPE western regional meeting, Irvine, California, 30 March–1 April.* https://doi.org/10.2118/93831-MS.

Seeton, C. J. (2006). Viscosity–temperature correlation for liquids. *Tribology Letters, 22*(1), 67–78. https://doi.org/10.1007/s11249-006-9071-2.

Weinbrandt, R. M., Ramey, H. J., Jr., & Casse, F. J. (1975). The effect of temperature on relative and absolute permeability of sandstones. *Society of Petroleum Engineers Journal, 15*(05), 376–384. https://doi.org/10.2118/4142-PA.

Abbreviations

AN	Acid number
BET	Brunauer-Emmett-Teller
BN	Base number
BPS	Bond product sum
CAPEX	Capital expenditures
CEC	Cation exchange capacity
CDC	Capillary desaturation curve
CGI	Continuous gas injection
CMC	Critical micelle concentration
CT	Computed tomography
CWI	Carbonated water injection
DPR	Disproportionate permeability reduction
EDL	Electrical double layer
EOR	Enhanced oil recovery
EOS	Equation of state
GC	Gas chromatography
HLB	Hydrophile-lipophile balance
HPAM	Hydrolyzed polyacrylamide
IAP	Ion activity product
ICP	Inductively coupled plasma
IOR	Improved oil recovery
LPG	Liquefied petroleum gas
LS	Low salinity
LSASPF	Low salinity-alternating-steam flood
LSPF	Low salinity–augmented polymer flood
LSWF	Low salinity waterflood
MICP	Mercury injection capillary pressure
MMC	Multiple mixing cell
MMP	Minimum miscibility pressure
NMR	Nuclear magnetic resonance
NPV	Net present value
OOIP	Original oil in place
OPEX	Operating expense
PALS	Phase-analysis-light-scattering
PET	Positron emission tomography
PNC	Pulsed neutron capture
PV	Pore volume
SAGD	Steam-assisted gravity drainage
SEM	Scanning electron microscope
SWAG	Simultaneous water and gas injection
SWCTT	Single well chemical tracer test
TAN	Total acid number
TDS	Total dissolved solids
TDT	Thermal decay time
TOC	Total organic carbon
TST	Transition state theory
WAG	Water-alternating-gas
WASP	Water-alternating-steam process
XRD	X-ray diffraction
XRF	X-ray fluorescence

Symbols

a_i and $[i]$	Activity of the species i
$\dot{a}_i$	Empirical ion-size parameter measuring effective diameter of the hydrated ion
e	Elementary charge
$f_{i,j}$	Fugacity of species i in phase j
g	Gravitational acceleration
i	Species
j	Phase
k	Permeability
k_a	Reaction rate constant
k_{rj}	Relative permeability of phase j
k_{rj}^{HS}	Relative permeability of phase j at the high threshold salinity condition
k_{rj}^{LS}	Relative permeability of phase j at the low threshold salinity condition
k_{rj}^{o}	Endpoint of relative permeability of phase j
$k_{rj}^{o\ HS}$	Endpoint of relative permeability of phase j at the high salinity threshold condition
$k_{rj}^{o\ LS}$	Endpoint of relative permeability of phase j at the low salinity threshold condition
$k_{salt,\ i}$	Salting-out coefficient of species i
k_B	Boltzmann constant
k_0	Reaction rate constant at reference temperature
m_i	Molality of species i
$meq_{i\text{-X}}$	Milliequivalent of the exchangeable species i
n_i	Number of moles of the species i
p	Pressure
p_c	Capillary pressure
p_c^{HS}	Capillary pressure at the high salinity threshold condition
p_c^{LS}	Capillary pressure at the low salinity threshold condition
$p_{H_2O}^{s}$	Saturation pressure of H_2O
r	Reaction rate
$\bar{v}_i$	Partial molar volume of species i
z_i	Charge of the ion
C_i	Equilibrium concentration of species i
$\widehat{C}_i$	Adsorbed concentration of species i
C_m	Mean curvature
CEC	Cation exchange capacity
E_a	Activation energy
F	Faraday constant
F_{IF}	Interpolation factor
G	Gibbs free energy
ΔG	Change in the Gibbs free energy
ΔG^o	Change in the standard Gibbs free energy of the reaction
$\overline{G}$	Effective molar Gibbs free energy of solution
$\overline{G}^{HS}$	Effective molar Gibbs free energy of solution at the high salinity threshold condition
$\overline{G}^{LS}$	Effective molar Gibbs free energy of solution at the low salinity threshold condition
H	Enthalpy
ΔH	Change in the enthalpy
H_i	Henry's constant of species i
H_i^s	Henry's constant of species i at the saturation pressure of H_2O, temperature, and zero salinity
$H_{salt,\ i}$	Henry's constant of species i in saline water
I	Ionic strength of the solution
IAP	Ionic activity product
K_a	Apparent dissociation constant
K_D	Partition coefficient
K_{eq}	Equilibrium constant
K_{int}	Intrinsic dissociation constant
K_{sp}	Solubility product
$K_{A/B}$	Exchange constant
$K'_{A/B}$	Selectivity coefficient
M	Mobility ratio
N_A	Avogadro constant
N_B	Bond number
N_c	Capillary number
N_T	Trapping number
R	Ideal gas constant
R_F	Resistance factor

R_k Permeability reduction factor
R_{RF} Residual resistance factor
S Entropy
ΔS Change in the entropy
S_o^* Normalized residual oil saturation
S_{or} Residual oil saturation
S_{or}^{high} Residual oil saturation at the high trapping number
S_{or}^{HS} Residual oil saturation at the high salinity threshold condition
S_{or}^{low} Residual oil saturation at the low trapping number
S_{or}^{LS} Residual oil saturation at the low salinity threshold condition
S_{wi} Irreducible water saturation
SI Saturation index
T Absolute temperature
$T_{r,\ H_2O}$ Reduced temperature of H_2O
U Internal energy
V Volume
W Colloidal interaction forces
γ_i Activity coefficient of species i
$\dot{\gamma}$ Shear rate
ε_r Dielectric constant
ε_0 Relative permittivity in free space
$\zeta(i\text{-}X)$ Equivalent fraction of the exchangeable species i
η Non-Newtonian fluid's viscosity
$[\eta]$ Intrinsic viscosity of non-Newtonian fluid
$\eta_{in-situ}$ In-situ viscosity of non-Newtonian fluid
η_I Inherent viscosity of non-Newtonian fluid
η_r Relative viscosity of non-Newtonian fluid
η_{sp} Specific viscosity of non-Newtonian fluid
η_0 Viscosity of non-Newtonian fluid at very low shear rate
η_∞ Viscosity of limiting value of non-Newtonian fluid at very high shear rate
θ Contact angle
θ^{HS} Contact angle at the high salinity threshold condition
θ^{LS} Contact angle at the low salinity threshold condition
θ^* Normalized contact angle between high and low salinity threshold conditions
κ^{-1} Debye length
λ Mobility
μ Viscosity
μ_i Chemical potential of the species i
μ_i^o Standard chemical potential of species i
ρ Density
σ Interfacial tension
ψ_0 Surface potential
$\prod(h)$ Disjoining pressure at the distance h between the oil and rock surfaces
Ω Saturation state

Index

Note: Page numbers followed by "f" indicate figures, "t" indicate tables.